THE ESSENTIALS OF CHEMICAL
THERMODYNAMICS

THE ESSENTIALS OF CHEMICAL THERMODYNAMICS

M. Kapel
B.Sc., Ph.D., CChem, MRSC, MIFST

Book Guild Publishing
Sussex, England

First published in Great Britain in 2011 by
The Book Guild Ltd
Pavilion View
19 New Road
Brighton, BN1 1UF

Typeset in Times by
YHT Ltd, London

Printed in Great Britain by
CPI Antony Rowe

A catalogue record for this book is available from
The British Library.

ISBN 978 1 84624 603 6

Contents

CONTENTS

Foreword

According to Edward Fitzgerald's translation, the eleventh century Persian poet Omar Khayyam described his early studies of philosophy with the words:

Myself when young did oft frequent
Doctor and Saint and heard great argument
About it and about, but evermore
Came out by the same door as in I went.

Unhappily, many chemistry students, after attending a lecture on thermodynamics, also come out 'by the same door as in they went'.

The subject of thermodynamics is widely acknowledged to be very important but notoriously difficult. Traditionally, it is approached in one of two ways, the classical and the statistical. The former of these presents to the student the various discoveries approximately in the order in which they were made. Although this may seem a logical procedure, the student is soon confronted with the conceptual difficulty of entropy, a quantity that is constantly increasing and appears to be continuously created out of nothing. By contrast, the statistical approach avoids this problem, but only at the expense of burdening the student with some mathematics of frightening complexity.

As a result of considerable experience in the teaching of thermodynamics, the author formed the conclusion that it is possible to broach the subject in such a way that its statistical nature can be made clear from the outset without recourse to the sort of mathematics that is only marginally accessible to the student but without sacrificing the rigorous nature of the argument. In this way, both the mathematical and conceptual difficulties are avoided.

Another obstacle to the student arises from the early introduction of free energy. When the concept of entropy has been incompletely mastered, that of free energy all too often serves only to add to the confusion. Moreover, the hapless novice is sometimes left with the erroneous impression that free energy is a fundamental product of the laws of nature, in the same way as energy or entropy, instead of a useful but not indispensable human artefact.

Yet another barrier confronting the student stems from the fact that many of the excellent books that exist on the subject were intended to function for both teaching and reference purposes. This renders such works extremely useful in many ways, but sometimes confuses the beginner, who finds it difficult to select the essentials from those details more appropriate for the adept.

Accordingly, the purpose of this book is firstly to explain the essentials of the subject in a way that is scientifically rigorous but avoids both the conceptual and mathematical difficulties that beset the more traditional approaches. Secondly, an attempt is made to explain the principal results of thermodynamic reasoning by considerations of entropy. Only when this has been achieved are the concepts of free energy and chemical potential introduced as an optional, though very useful, alternative technique. Finally, the author has taken care to limit the material strictly to the essentials, so that the reader may come to a firm understanding of these before widening his or her knowledge by referring to more advanced treatises.

Since the many symbols used in thermodynamics can often be confused with those denoting units, the latter are treated separately on a page at the beginning of the book, and are avoided in the text itself except when attention needs to be drawn to them for a particular reason.

Acknowledgments do not usually find a place in the foreword of a book, being more commonly reserved for a separate section. Nevertheless, this seems to be the most appropriate place to record my indebtedness to the late Mr J.E.B. Randles, whose lectures to undergraduates, delivered more than half a century ago, first kindled my interest in the subject, and the clarity of whose explanations ensured that I never 'came out by the same door as in I went'.

Quantities and Units

Intensive and Extensive Quantities

The quantities used in thermodynamics can be classified into two categories, namely intensive and extensive ones. The value of an intensive quantity does not depend on the amount of substance being considered, a good example being temperature. By contrast, the value of an extensive quantity increases when extra material is added. A good example of an extensive quantity is volume. When an intensive quantity is multiplied by an extensive one, the product is always extensive.

The SI units of the quantities and constants used in this book are given below.

Heat	Joules	J
Work	Joules	J
Pressure	Newtons per square metre	$N\ m^{-2}$
Volume	Cubic metres	m^3
Molar volume	Cubic metres per mole	$m^3\ mol^{-1}$
Internal energy	Joules	J
Molar internal energy	Joules per mole	$J\ mol^{-1}$
Enthalpy	Joules	J
Molar enthalpy	Joules per mole	$J\ mol^{-1}$
Area	Square metres	m^2
Distance	metres	m
Temperature	Kelvin	K
Molar thermal capacity	Joules per mole per Kelvin	$J\ mol^{-1}\ K^{-1}$

Gas constant	Joules per mole per Kelvin	$J\ mol^{-1}\ K^{-1}$
Avogadro number	Molecules per mole	mol^{-1}
Boltzmann constant	Joules per molecule per Kelvin	$J\ molecule^{-1}\ K^{-1}$
Entropy	Joules per Kelvin	$J\ K^{-1}$
Molar entropy	Joules per mole per Kelvin	$J\ mol^{-1}\ K^{-1}$
Concentration	Moles per cubic metre	$mol\ m^{-3}$
Molar latent heat	Joules per mole	$J\ mol^{-1}$
Fugacity	Newtons per square metre	$N\ m^{-2}$
Activity	Moles per cubic metre	$mol\ m^{-3}$
Faraday	Coulombs per mole	$C\ mol^{-1}$
Potential	Volts	V
Standard electrode potential	Volts	V
Redox potential	Volts	V
Gibbs Free Energy	Joules	J
Helmholtz Free Energy	Joules	J
Chemical potential	Joules per mole	$J\ mol^{-1}$

1

The Heart of the Problem

Most historians would agree that the momentous political events that occurred in Paris in 1789 changed the complexion of the whole world. It is less widely known, however, that these stirring happenings completely overshadowed a scientific discovery which, occurring in the same year and in the same city as the French Revolution, was to have no less profound an effect on the future of mankind, for it was in 1789 that Antoine Lavoisier, after years of painstaking experiments, at last articulated the conclusion that, in any chemical reaction, the mass of the products must equal that of the reactants. From this, he concluded that the total amount of matter in the universe was constant, and that nothing material could be either created or destroyed.

Since Lavoisier's time, the Law of Conservation of Matter, as the proposition came to be called, has required some modification. In 1905, Einstein discovered that matter and energy were different manifestations of the same thing. A Law of Conservation of Energy had already been propounded by R.J. von Mayer in 1842[1], so that henceforth what was considered to be conserved was neither matter nor energy, but rather the sum total of the two. Moreover, the considerations that led Einstein to his conclusion served to warn of the dangers of extrapolation from the results obtained within the limited confines of a laboratory and finite duration of an experiment to the enunciation of a theory concerned with the vastness of the universe over a very long period of time. Thus, if matter were being created or destroyed at a slow rate throughout the universe, the extent of the change in those small elements of space and time

---

[1] R.J. von Mayer had some difficulty in persuading the scientific community of the validity of his views. When the Law of Conservation of Energy was propounded independently two years later by H. von Helmholtz, it enjoyed a better reception, but its real importance does not appear to have been appreciated before the work of J.P. Joule.

occupied by an experiment might not be measurable by means of the techniques currently available. One cannot, therefore, regard the Law of Conservation of Matter and Energy as definitely proved, but, whatever its theoretical status, the Law must be considered to be either true or so close an approximation to the truth that it may be deemed valid for all practical purposes. In most chemical experiments, there is some liberation or absorption of energy, which must presumably be accompanied by a change in the mass of the system. It is true to say, however, that the quantities of energy normally involved would be equivalent to such minute changes of mass that the latter would be undetectable. Accordingly, the Law of Conservation of Matter, as enunciated by Lavoisier, may be regarded as true for the purposes of the practical chemist, and any experimental predictions based on the Law, if otherwise valid, should be capable of fulfilment in the laboratory.

If nothing can be created or destroyed during an experiment, one may ask oneself what is achieved in a chemical reaction. For this purpose, let us consider a typical chemical change:

$$CaCO_3 + 2HCl \rightarrow CaCl_2 + H_2O + CO_2.$$

The reagents on the left consist of a certain number of particles of various kinds. According to the Law of Conservation of Matter, the products on the right must consist of exactly the same number of particles of the same kinds. One is, therefore, led to the inescapable conclusion that the only difference between the reagents and the products lies in the arrangement of the particles. A chemical reaction may thus be regarded as a re-arrangement.

This conclusion immediately raises the question why the substances given on the left of the equation are the reagents and those on the right are the products, in short, why the re-arrangement can occur in one direction and not in the other. To be sure, there are reversible reactions such as, for example, the following:

$$CaCO_3 \rightleftarrows CaO + CO_2.$$

Nevertheless, many reactions are quite indubitably irreversible.

It would seem, therefore, that we have to consider two distinct types of re-arrangement, namely those that can occur in one direction only, and those that can proceed in both directions.

2

A moment's reflection will make us realise that there is yet a third type of re-arrangement. This is typified by the three formulae below:

$$CH_3.CONH_2 \qquad CH_3.CH:N.OH \qquad CH_3.CH_2.NO.$$

These three substances are isomeric, differing only in the arrangement of the atoms in the molecules, yet it does not appear to be possible to convert any one of these compounds directly into one of the others. Hence, the third type of re-arrangement is one that can be postulated in theory, but cannot seemingly be realised in practice. If an explanation is required for the existence of the three types of chemical equation – representing the reactions that are irreversible, those that are reversible, and those that are theoretically possible but practically unattainable – then it must clearly be sought in the theory of arrangements, with which we must now concern ourselves.

Let us consider an ordinary pack of playing cards, and let us isolate the 13 cards of one suit. Let the cards be stacked in ascending order of value. The result is an orderly arrangement, but, if the 13 cards are now shuffled, the tidiness of the arrangement will quickly be destroyed. Nevertheless, if the shuffling is continued for a sufficiently long period, then it is quite possible for the original order to be fortuitously re-established. The length of time that this will take is determined by the vagaries of chance, but, if the rate of shuffling is specified, an average period for the recurrence of the original order may be calculated as explained below.

It is possible for any one of the 13 cards to be on top at any stage of the shuffling process. For each one of the 13 ways of selecting the top card, there are then 12 ways of selecting the second, so that the top two cards can appear in any one of 13×12 different ways. For each of these possibilities, there are 11 ways in which the third card may be chosen, and, accordingly, there are $13 \times 12 \times 11$ distinct arrangements for the first three cards. An extension of this argument gives the total number of arrangements for all 13 cards as

$$13 \times 12 \times 11 \times 10 \times 9 \times 8 \times 7 \times 6 \times 5 \times 4 \times 3 \times 2 \times 1.$$

This number is usually referred to as "factorial 13" and written 13!. Since, during the shuffling, any one of these arrangements is as likely to occur as any other, it follows that the original order will re-assert itself on the average in one out of every 13! shuffling movements. If we stipulate that the shuffling is to consist of one movement per second, then we may expect to encounter the original order once in every

$$13 \times 12 \times 11 \times 10 \times 9 \times 8 \times 7 \times 6 \times 5 \times 4 \times 3 \times 2 \times 1$$
seconds.

This period may be converted to more manageable units as follows:-

To convert to minutes, divide by 60 by deleting 6 and 10.
To convert to hours, divide by 60 by deleting 5 and 12.
To convert to days, divide by 24 by deleting 3 and 8.
To convert to weeks, divide by 7 by deleting 7.
To convert to years, divide by 52 by deleting 4 and 13.

It follows that the period in question is

$$11 \times 9 \times 2 \times 1 = 198 \text{ years.}$$

Since a year is a little longer than 52 weeks, a more accurate figure is slightly more than 197 years.

One might suppose that, since the period for 13 cards is almost 200 years, that for the whole pack would be four times as long, i.e. 800 years, but such reasoning is fallacious. Even if only one card were added to the original 13, the period would be 14! seconds, or 14 times as long as in the case already discussed. The addition of one further card would multiply this new period by 15. The results of such additions are given in Table 1.1.

Table 1.1 Average period for recurrence of original order in a pack of cards during the shuffling process

Number of Cards	Average Period for Re-appearance of Order
13	197 years
14	2,763 years
15	41,439 years
16	663,017 years
17	11,271,286 years
18	202,883,146 years
19	3,854,779,777 years
20	77,095,595,549 years
52	ca. 2.556×10^{60} years

4

The most striking features of these figures are not merely their tremendous magnitude, but also their very high, and rapidly accelerating, rate of increase. Even so small a number of cards as 52 is associated with a period of time beyond the scope of the human imagination, whilst the addition of only one further card would require the number to be multiplied by 53.

Molecules or ions, however, are not normally encountered in numbers as small as 52. Indeed, in one single burette drop of about 0.05 cm^3, there are approximately 1.67×10^{21} molecules of water, all constantly being shuffled as a result of their kinetic motion. Moreover, the re-establishment of any particular order must be concerned not only with the geometrical arrangement of the molecules, but also with the distribution of energy among them and with the allocation of this energy among the various translational, rotational, vibrational and electronic degrees of freedom. Even when account is taken of the fact that each shuffling movement takes very much less than one second (10^{-12} second is probably a more realistic estimate), the fortuitous re-establishment of any particular arrangement can be seen to be an event of great rarity. With a number of molecules as low as 100, the average period for the reproduction of a geometrical sequence only would be 2.96×10^{138} years, an interval many times as great as the estimated age of the universe. With the numbers of molecules and ions normally encountered in a reaction, the corresponding period would be so great, that the reappearance of an ordered arrangement must be considered to be, for all practical purposes, impossible. The key to the mechanism of all chemical and physical changes is, therefore, to be found in an associated increase in the disorder of the molecular system. A return to order is not impossible, but is an event of such low probability, that the possibility of its occurrence may be discounted.

It is at this point that the reader should be warned of a fallacy that not infrequently besets thermodynamic reasoning. As pointed out above, the shuffling of 13 cards at the rate of one movement per second results in the recurrence of the original order only about once in every 197 years. If, however, only two of the 13 cards are considered, then the re-establishment of a sequence will occur very much more frequently. In the same way, in any system in which a chemical reaction occurs, order may well be restored in a part of the assembly of molecules while a compensating increase in disorder occurs elsewhere. Thus, if the criterion of increasing disorder is to be applied, it must involve the whole of the system comprising not only the reagents and products themselves, but also the molecules of the environment capable of absorbing or supplying the energy liberated or required by the reaction mixture. The exchange of energy between the reaction mixture and its environment is clearly of crucial importance, since the supply or withdrawal of energy will affect the kinetic motion of molecules and hence their degree of randomness.

One objection to the line of argument detailed above is its theoretical nature. While some understanding of the driving force of a chemical reaction or physical change may indeed be provided by these considerations, it would be highly desirable to be able to convert the theoretically calculable disorder of a system into some practically measurable criterion. The clue to this problem has already been mentioned. Since the exchange of energy between a reaction mixture and its environment exerts an effect on the degree of disorder of the latter, it may well serve as a measure of the change in disorder occurring in the system as a whole, and may thus be looked upon as an experimentally determinable indicator of the statistical progress of the change. It will, however, be necessary to find some way of correlating absorption or evolution of energy with changes in the degree of disorder. In the case of a solid, this is very difficult, since the supply of energy does not simply increase the kinetic energy of translational motion owing to the complicated intermolecular forces existing in a crystal lattice. In lesser degree, the same objection applies to the study of liquids, and, even in the realm of real gases, intermolecular forces exist to complicate the issue. The study of thermodynamics must, therefore, begin with a consideration of ideal gases, and it is this, as well as more general considerations of energy change, that will occupy the early chapters of this book.

2

Heat and Mechanical Energy

The First Law of Thermodynamics

One of the consequences of the Law of Conservation of Energy is the proposition that has become known as the First Law of Thermodynamics. This may be enunciated as follows:

Heat energy and mechanical energy are interchangeable, and a given quantity of the one may be converted into the equivalent quantity of the other.

In its original form, the Law contained another clause, for it was formerly customary to measure heat and mechanical energy in different units. The conversion factor was accordingly contained in the original statement of the Law, but the emergence of the SI system of units, in which both heat and mechanical energy are measured in joules, has rendered the so-called "mechanical equivalent of heat" unnecessary.

In order that the First Law may be applied to a study of the interrelationship of heat and mechanical work, certain symbols must first be defined:-

Let the heat absorbed by a system $= q$.
Let the work done by the system $= w$.
Let the pressure acting on the system $= P$.
Let the volume occupied by the system $= V$.

A few points should be made about these symbols. Firstly, if the quantities of heat and work are infinitesimally small, they are usually represented by dq and dw respectively in accordance with normal mathematical practice. Secondly, it is necessary to draw a clear distinction between work which is done by the system and that done on the system. For our present purpose, work done by the system will be regarded as positive in sign and that done on the system as negative.

It is now possible to define two further quantities – internal energy, and enthalpy or heat content.

Internal Energy (U)

The internal energy of a system is made up of kinetic energy, various forms of potential energy and possibly some other components as well. A consideration of the potential energy alone will serve to illustrate the difficulty of assigning a numerical value to the quantity U. If an object is held at some distance above the ground, it will possess potential energy in virtue of its gravitational attraction to the earth. The magnitude of this energy will depend upon the weight of the object and its distance above the ground. The latter, however, can be altered by the simple process of digging a hole at the point over which the object is suspended. Thus, the potential energy of the object can seemingly be modified by a process that is in no way connected with the object itself. If, however, two objects had been suspended at different heights over the same point, then the digging of a hole, while altering the potential energy of each, would leave the difference between them unchanged. It therefore follows that no absolute value can ever be assigned to potential energy, but that the difference between two potential energies can be measured precisely. Since potential energy is a component of total energy, the same comment must apply to this also.

The First Law of Thermodynamics informs us that an absorption of heat must result in an increase in internal energy, while the performance of work by a system must occasion a decrease. Accordingly, a change in the internal energy of a system may be defined by the equation

$$dU = dq - dw.$$

Enthalpy or Heat Content (H)

The enthalpy, or heat content, of a system is defined as follows:-

$$H = U + PV.$$

8

Since it is not possible to state an absolute value for internal energy, the same must be true of enthalpy, so that only changes in this quantity have any fundamental meaning.

Although enthalpy is often referred to as heat content, its connection with heat is not immediately obvious from the definition given above. Nevertheless, the term "heat content" is apt for reasons that will presently be explained.

Work Done by an Expanding System

Figure 2.1

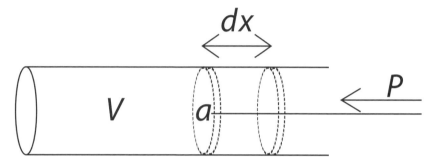

Consider a volume V of gas confined in a cylinder by a piston having an area of cross section a and exerting a pressure P. Let the temperature of this gas be raised slightly. The ensuing expansion will force the piston back by a distance dx. The volume will, therefore, increase by an amount equal to the volume of the small cylindrical element contained between the initial and final positions of the piston (Figure 2.1). Thus,

$$\text{increase in volume} = dV = a.dx.$$

The piston was, however, exerting a force on the gas, and the expansion of the latter has caused the force to move its point of application. According to the normal principles of mechanics, this means that work has been done. The following relationships may thus be stated:-

Force exerted on gas = pressure × area of piston = Pa.
Work done by gas = force × distance = $Pa.dx$ = $P.dV$.

Although it was easier to illustrate this point by means of a gas, it will be noted that the properties of gases have not been used at any stage in the argument. The final result must, therefore, be equally true of solids and liquids. In the context of any particular problem, it may be possible to equate the quantity $P.dV$ with some other expression and so to obtain results of particular interest or application. Under all circumstances, however, it will be true to say that the work done on expansion against an externally applied pressure is equal to $P.dV$, and this statement serves as the starting point of many calculations.

Heat Absorbed by a System

The absorption of heat by a system engenders a tendency to expansion, and, if the pressure on the system is kept constant, an increase in volume will occur. Alternatively, the expansion may be checked by a gradual increase in the externally applied pressure, so that the volume is kept constant. Accordingly, the effects of heat absorption must be considered under two distinct sets of conditions, namely constant pressure and constant volume. The second of these will be discussed first.

Constant Volume

Under conditions of constant volume, the following mathematical relationships hold:-

$$V = \text{constant.}$$
$$dV = 0.$$
$$dw = P.dV = 0.$$
$$dq = dq - dw = dU.$$
$$q = \Delta U.$$

Thus, the heat absorbed by a system at constant volume is equal to the increase in internal energy.

Constant Pressure

Under conditions of constant pressure, we may argue as follows:-

$$P = \text{constant.}$$
$$dw = P.dV = d(PV).$$
$$dq = (dq - dw) + dw$$
$$= dU + d(PV)$$
$$= d(U + PV)$$
$$= dH.$$
$$q = \Delta H.$$

This means that the heat absorbed by a system under constant pressure is equal to the increase in enthalpy. The reason for the alternative name of enthalpy should now be clear.

Actually, an important limitation exists for both of these arguments. In each case, it was assumed that the only work done was that due to expansion against the externally applied pressure. In some systems, however, this is not the case, and electrical or other work serves to complicate the issue. In such cases, the expressions for q must be suitably modified.

The Molar Thermal Capacity

The amount of heat required to raise the temperature of one mole of a substance by 1K is referred to as the molar thermal capacity of that substance. It is not a constant quantity, but depends on the magnitude of the starting temperature, and also on whether the volume or pressure is kept constant. If C_V and C_P are used to denote the molar thermal capacities at constant volume and pressure respectively, then the following equations may be written:-

$$C_V = (q/\Delta T)_V.$$
$$C_P = (q/\Delta T)_P.$$

(The subscripts V and P denote constant volume and constant pressure respectively.)

Although, as already stated, the quantities C_V and C_P vary according to the initial temperature, little error results if they are regarded as constant over the range of one degree. In this case, the two functions may be rewritten as follows:-

$$C_v = \left(\frac{\partial U}{\partial T}\right)_v.$$

$$C_P = \left(\frac{\partial H}{\partial T}\right)_P.$$

The symbol ∂ is that of partial differentiation, a mathematical technique that is particularly useful in thermodynamics. For the benefit of the reader unfamiliar with this procedure, it is explained in the Appendix.

It is possible to deduce a relationship between C_V and C_P by the following reasoning. The temperature, volume and pressure of one mole of a substance are related, so that a specification of values for any two of them automatically fixes the third. The internal energy is, therefore, defined if any two of them, say the volume and temperature, are stipulated. In accordance with the normal rules of partial differentiation, an increase in internal energy may be written as follows:-

$$dU = \left(\frac{\partial U}{\partial T}\right)_V.dT + \left(\frac{\partial U}{\partial V}\right)_T.dV$$

$$= C_V.dT + \left(\frac{\partial U}{\partial V}\right)_T.dV.$$

Hence,

$$\left(\frac{\partial U}{\partial T}\right)_P = C_V + \left(\frac{\partial U}{\partial V}\right)_T.\left(\frac{\partial V}{\partial T}\right)_P.$$

Now

$$H = U + PV$$

and

$$C_P = \left(\frac{\partial H}{\partial T}\right)_P.$$

Therefore,

$$C_P = \left(\frac{\partial U}{\partial T}\right)_P + P\left(\frac{\partial V}{\partial T}\right)_P.$$

$$C_P = C_V + \left[P + \left(\frac{\partial U}{\partial V}\right)_T\right]\left(\frac{\partial V}{\partial T}\right)_P.$$

At first sight, this equation may seem too complicated to be possessed of any real usefulness, but it will be demonstrated later that, in certain special cases, notably that of an ideal gas, the expression for C_P can be simplified into something that is easily calculable.

In the meantime, the true significance of the relationship should be understood. This is best accomplished by a multiplication of the equation by dT to give

$$C_P.dT = C_V.dT + P.dV + dU.$$

The individual terms may now be explained:-

$C_P.dT$ = heat required to produce change of temperature by dT at constant pressure.
$C_V.dT$ = heat required to cause actual increase in temperature, i.e. change in kinetic energy of molecules.
$P.dV$ = work of expansion against externally applied pressure.
dU = work of expansion against intermolecular forces.

The interpretation of dU is clear from its origin in the equation for C_P, since the term

$$\left(\frac{\partial U}{\partial V}\right)_T$$

represents a change in internal energy resulting from a change in

13

volume at constant temperature. The latter condition, however, requires the kinetic energy of the molecules to be constant, so that the change must be one of potential energy. This can only occur if the molecules moving apart are bound by some intermolecular forces.

3

The Properties of Ideal Gases

The Description of an Ideal Gas

As is well known, the gaseous state is a condition in which the molecules of a substance are free to move about the containing vessel without the constraints normally present in solids and liquids. It is, however, an oversimplification to suppose that the only interaction between them occurs as a result of chance collisions, for intermolecular forces, though very much less than in the other two states of matter, are never entirely absent. Nevertheless, it is possible to postulate an ideal substance in which the molecules enjoy complete independence of one another, and to draw conclusions concerning the effects of temperature and pressure changes on such a material. Even the most elementary considerations lead to one obvious conclusion, namely that any calculation of volume change must take into account two factors, the space between the molecules and the molecules themselves. The former of these is easily amenable to change, but the molecules themselves are virtually inflexible. This unwelcome complication is easily removed by the further postulation of such conditions of temperature and pressure, that the molecules are well separated and contribute only a minute proportion of the total volume. In short, a theoretically assumed state in which the molecules are sufficiently far apart to form a negligible fraction of the total volume and to be immune to intermolecular forces of attraction and repulsion is known as an ideal gas, and, although no real gas can comply absolutely with these stipulations, there are some that, at sufficiently high temperature and low pressure, closely approach the ideal state.[1]

---

[1] Another requirement for an ideal gas is that collisions between molecules must be perfectly elastic, but, important though this is in other contexts, it plays no part in the arguments developed in this chapter.

Theoretical considerations lead to the following equation for an ideal gas:

$$PV = RT, \qquad (3.1)$$

where

P = pressure,
V = volume occupied by one mole,
R = gas constant,
and T = absolute temperature.

This equation is commonly known as the ideal gas equation.

The absence of intermolecular forces in an ideal gas means that an expansion of the gas, i.e. an increase in the intermolecular distances, does not result in any change in potential energy. Since the kinetic energy of the molecules is a function of the temperature only, it follows that an expansion at constant temperature does not change the internal energy of the system. Accordingly,

$$\left(\frac{\partial U}{\partial V}\right)_T = 0. \qquad (3.2)$$

If each side of equation (3.1) is differentiated with respect to T at constant pressure, the following result will be obtained:

$$P\left(\frac{\partial V}{\partial T}\right)_P = R. \qquad (3.3)$$

In Chapter 2, a relationship between C_P and C_V was established. According to this,

$$C_P = C_V + \left[P + \left(\frac{\partial U}{\partial V}\right)_T\right]\left(\frac{\partial V}{\partial T}\right)_P.$$

The application of equations (3.2) and (3.3) now leads to the following result, which is valid for ideal gases:

$$C_P = C_V + P\left(\frac{\partial V}{\partial T}\right)_P$$

$$= C_V + R.$$

Hence,

$$C_P - C_V = R. \qquad (3.4)$$

This important result will feature prominently in the arguments detailed below.

Equation (3.1) contains three variable quantities, namely pressure, volume and temperature, and it is possible to alter any two of these while the third is kept constant. This means that any discussion of the properties of ideal gases must include a description of these three types of change. A fourth type, the adiabatic change, must also be considered.

Changes at Constant Pressure

If the pressure acting on a gas is kept constant, the ideal gas equation becomes simplified to the well-known Charles's Law:-

$$P = \text{constant.}$$
$$PV = RT.$$
Therefore, $V \propto T$ (Charles's Law).

Let the temperature of the gas be raised from T_1 to T_2, and the volume of one mole be correspondingly increased from V_1 to V_2. Then the work done by the expanding gas may be calculated as follows:-

$$w = \int_{V_1}^{V_2} P.dV$$

$$= P(V_2 - V_1).$$

The application of the ideal gas equation gives an alternative expression:

$$w = R(T_2 - T_1).$$

The heat absorbed by one mole of the gas may also be calculated:

$$q = \int_{T_1}^{T_2} C_P.\mathrm{d}T.$$

The evaluation of the integral is complicated by the dependence of C_P on the temperature. If, however, the temperature change is not too great, then C_P may be regarded as constant, in which case

$$q = C_P(T_2 - T_1).$$

It is now possible to deduce an expression for the change in internal energy:-

$$\Delta U = q - w$$
$$= C_P(T_2 - T_1) - R(T_2 - T_1)$$
$$= (C_P - R)(T_2 - T_1).$$

The application of equation (3.4) changes this to

$$\Delta U = C_V(T_2 - T_1).$$

Our final consideration is the change in enthalpy.

$$\Delta H = \Delta(U + PV)$$
$$= \Delta U + \Delta(PV)$$
$$= \Delta U + \Delta(RT)$$
$$= C_V(T_2 - T_1) + \int_{T_1}^{T_2} R.\mathrm{d}T$$
$$= C_V(T_2 - T_1) + R(T_2 - T_1)$$
$$= (C_V + R)(T_2 - T_1)$$
$$= C_P(T_2 - T_1).$$

It will be noticed that, for a change at constant pressure,

$$\Delta H = q.$$

This is in accordance with the discussion of q in Chapter 2.

Changes at Constant Volume

At constant volume, the ideal gas equation may again be simplified, this time into a proportional relationship between the pressure and the absolute temperature:-

$$V = \text{constant.}$$
$$PV = RT.$$
$$\text{Therefore, } P \propto T.$$

The quantities w, q, ΔU and ΔH can be calculated with the following results:-

$$V = \text{constant.}$$
$$dV = 0.$$
$$dw = P.dV = 0.$$
$$w = 0.$$

$$q = \int_{T_1}^{T_2} C_V.dT$$

$$= C_V(T_2 - T_1) \text{ (if the temperature difference is not too great).}$$

$$\Delta U = q - w$$
$$= q$$
$$= C_V(T_2 - T_1).$$

$$\Delta H = \Delta(U + PV)$$
$$= \Delta U + \Delta(PV)$$
$$= \Delta U + \Delta(RT)$$

$$= C_V(T_2 - T_1) + \int_{T_1}^{T_2} R.dT$$

$$= C_V(T_2 - T_1) + R(T_2 - T_1)$$
$$= (C_V + R)(T_2 - T_1)$$
$$= C_P(T_2 - T_1).$$

Once again, it may be noted that the equality of q and ΔU is in agreement with the considerations given in Chapter 2.

Isothermal Changes

Changes in which the temperature is kept constant are of particular importance, and are known as isothermal changes. They are amenable to calculations of a kind similar to those already performed for the other two instances.

$$T = \text{constant.}$$
$$PV = RT.$$
$$\text{Therefore, } PV = \text{constant (Boyle's Law).}$$

$$w = \int_{V_1}^{V_2} P.dV.$$

The integration of this expression occasions a slight difficulty, since P varies continuously as V changes. Since, however, T is constant, the ideal gas equation may be invoked.

$$w = \int_{V_1}^{V_2} P.dV$$

$$= \int_{V_1}^{V_2} \frac{RT}{V}.dV$$

$$= RT \ln(V_2/V_1).$$

This is one possible expression for w, but the application of Boyle's Law yields another.

$$w = RT \ln(V_2/V_1) = RT \ln(P_1/P_2).$$

Since there is no change of temperature, the heat absorbed must correspond exactly with the external work done, so that

$$q = RT \ln(V_2/V_1) = RT \ln(P_1/P_2).$$

It now follows that

$$\Delta U = q - w = 0,$$

and that

$$\Delta H = \Delta(U + PV)$$
$$= \Delta U + \Delta(PV)$$
$$= 0 \text{ (since } PV \text{ is constant).}$$

Adiabatic Changes

Another process of great importance is the adiabatic change. In this, the pressure, volume and temperature all change, but in such a way that no heat is either absorbed or evolved by the gas. Accordingly, the adiabatic change may be defined by the equation

$$q = 0.$$

The ideal gas equation still holds good, as does the equation

$$C_P - C_V = R,$$

but this time it is more difficult to derive an equation that correlates the changes in two of the variables. Indeed, since all three properties change, three such equations will exist.

As before, let us consider the pressure to change from P_1 to P_2, the volume of one mole from V_1 to V_2 and the temperature from T_1 to T_2. Let us also introduce a quantity γ defined by the equation

$$\gamma = C_P/C_V.$$

The change in internal energy may now be calculated in two ways. From the definition given in Chapter 2, it follows that

$$
\begin{aligned}
\mathrm{d}U &= \mathrm{d}q - \mathrm{d}w \\
&= -\,\mathrm{d}w \\
&= -\,P.\mathrm{d}V \\
&= -\,(RT/V).\mathrm{d}V \\
&= -\,(C_P - C_V)(T/V).\mathrm{d}V.
\end{aligned}
$$

The other way of calculating $\mathrm{d}U$ is derived from the rules of partial differentiation. If the volume and temperature are regarded as independent variables, then

$$
\mathrm{d}U = \left(\frac{\partial U}{\partial V}\right)_T .\mathrm{d}V + \left(\frac{\partial U}{\partial T}\right)_V .\mathrm{d}T.
$$

For an ideal gas, however,

$$
\left(\frac{\partial U}{\partial V}\right)_T = 0 \quad \text{(see equation (3.2))}.
$$

Thus,

$$
\mathrm{d}U = \left(\frac{\partial U}{\partial T}\right)_V .\mathrm{d}T
$$

$$
= C_V.\mathrm{d}T.
$$

The two expressions for $\mathrm{d}U$ may now be equated, so that

$$
C_V.\mathrm{d}T = -\,(C_P - C_V)(T/V).\mathrm{d}V.
$$

Hence,

$$
\int_{T_1}^{T_2}\frac{\mathrm{d}T}{T} = \int_{V_1}^{V_2} -(\gamma-1).\frac{\mathrm{d}V}{V}.
$$

$$
\ln(T_2/T_1) = \ln(V_1/V_2)^{(\gamma-1)}.
$$

22

$$T_2/T_1 = (V_1/V_2)^{(\gamma-1)}.$$
$$T_1 V_1^{(\gamma-1)} = T_2 V_2^{(\gamma-1)}.$$

In general, it may be said that

$$TV^{(\gamma-1)} = \text{constant.} \qquad (3.5)$$

This equation relates the temperature of the gas to the volume occupied by one mole. An expression involving pressure and volume may easily be derived by means of the substitution

$$T = PV/R.$$

From this, it follows that

$$PV^{\gamma} = \text{constant,} \qquad (3.6)$$

the quantity R having become incorporated in the constant on the right-hand side. By means of the substitution

$$V = RT/P,$$

the equation expressing the change of temperature with pressure is found to be

$$T^{\gamma}/P^{(\gamma-1)} = \text{constant.} \qquad (3.7)$$

The equations (3.5), (3.6) and (3.7) express the interdependence of the pressure, temperature and volume occupied by a given quantity of an ideal gas during an adiabatic change. The values of w, q, ΔU and ΔH for the adiabatic expansion of one mole may now be calculated as follows.

It was shown above that

$$dU = -dw,$$

and that

$$dU = C_V.dT.$$

Accordingly,

$$dw = -C_V.dT,$$

so that

$$w = \int_{T_1}^{T_2} -C_v.dT$$

$$= C_V(T_1 - T_2).$$

As before, the value of the integral is subject to the limitation that T_1 and T_2 must be sufficiently close together for C_V to be effectively constant over the temperature range.

The value of q is enshrined in the definition of an adiabatic change, according to which

$$q = 0.$$

It may now be seen that

$$\Delta U = q - w$$
$$= C_V(T_2 - T_1).$$

Finally, the value of ΔH remains to be calculated.

$$\Delta H = \Delta(U + PV)$$
$$= \Delta U + \Delta(PV)$$
$$= \Delta U + \Delta(RT)$$
$$= C_V(T_2 - T_1) + \int_{T_1}^{T_2} R.dT$$
$$= C_V(T_2 - T_1) + R(T_2 - T_1)$$
$$= (C_V + R)(T_2 - T_1)$$
$$= C_P(T_2 - T_1).$$

Summary

Table 3.1 The behaviour of an ideal gas

	Constant Pressure	Constant Volume	Isothermal Change	Adiabatic Change
	V/T = constant	P/T = constant	PV = constant	$TV^{(\gamma-1)}$ = constant
				PV^{γ} = constant
				$T^{\gamma}/P^{(\gamma-1)}$ = constant
w	$P(V_2 - V_1)$	0	$RT \ln(V_2/V_1)$	$C_V(T_1 - T_2)$
	$= R(T_2 - T_1)$		$= RT \ln(P_1/P_2)$	
q	$C_P(T_2 - T_1)$	$C_V(T_2 - T_1)$	$RT \ln(V_2/V_1)$	0
			$= RT \ln(P_1/P_2)$	
ΔU	$C_V(T_2 - T_1)$	$C_V(T_2 - T_1) = q$	0	$C_V(T_2 - T_1)$
ΔH	$C_P(T_2 - T_1) = q$	$C_P(T_2 - T_1)$	0	$C_P(T_2 - T_1)$

The behaviour of an ideal gas may be summarised as shown in Table 3.1. It will be observed that, in all cases,

$$\Delta U = C_V(T_2 - T_1)$$
$$\text{and } \Delta H = C_P(T_2 - T_1)$$

though, for isothermal changes, both of these expressions can obviously be equated with zero.

Finally, it is proved in textbooks of physical chemistry that γ is equal to 1.67 for monatomic gases and 1.4 for those having linear molecules. The latter category includes all diatomic gases and a few others besides. Those molecules possessing three or more atoms not disposed in a straight line should ideally give rise to values of 1.33 for the quantity γ, but, the more complicated the structure, the greater is the departure from this theoretical figure.

4

Heat, Temperature and Disorder

Ideal Gases

Figure 4.1

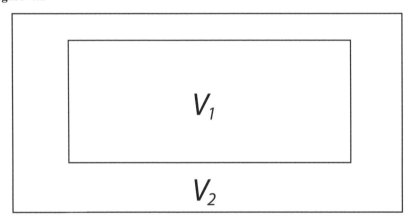

Let us consider a volume V_2, within which a smaller volume V_1 is defined. It is important to remember that V_2 represents the total volume inclusive of V_1 (Figure 4.1).

Let a molecule of a gas be introduced into the system. The number of ways in which this molecule can be positioned is clearly proportional to the volume within which it is to be found, so that the following relationships may be deduced.

Let the number of ways in which the molecule can occupy volume $V_1 = W_1$.
Let the number of ways in which the molecule can occupy volume $V_2 = W_2$.

Then

$$W_1 = aV_1 \text{ (where } a \text{ is a suitable constant),}$$

and

$$W_2 = aV_2.$$

Thus,

$$\frac{W_2}{W_1} = \frac{V_2}{V_1}.$$

Let us now consider two molecules of the gas to be involved. There are still aV_1 ways in which one of these can occupy volume V_1, but, for each of these different possibilities, there are also aV_1 ways in which the second molecule can be found there, since, as was stated earlier, the molecules of an ideal gas are considered to be infinitesimally small. Thus, if W_1 now represents the number of ways in which both molecules can be found in the small inner compartment, then

$$W_1 = a^2V_1{}^2.$$

By similar reasoning, if W_2 denotes the number of ways in which both molecules can arrange themselves throughout the whole system,

$$W_2 = a^2V_2{}^2.$$

Accordingly,

$$\frac{W_2}{W_1} = \left(\frac{V_2}{V_1}\right)^2.$$

This line of reasoning can be extended to any other number, N, of molecules, in which case,

$$\frac{W_2}{W_1} = \left(\frac{V_2}{V_1}\right)^N. \tag{4.1}$$

If we consider one mole of gas to be introduced, then equation (4.1) remains true, but N now assumes the special meaning of the number of molecules in one mole, i.e. the Avogadro number.

It is clear that the quantities W_1 and W_2 are very large numbers, and that any arithmetical manipulations of them might well be complicated by this fact. Such an inconvenience is, however, readily avoided by the use not of the W's themselves, but of some logarithmic function. For this reason, it is expedient to define a new quantity, S, so that

$$S = k \ln W,$$

where the logarithm is to the base e and k is given by the equation

$$k = \frac{R}{N},$$

R being the gas constant and N the Avogadro number. The quantity k is deemed to be of sufficient importance to merit a name, and is accordingly called the Boltzmann constant.

If one mole of an ideal gas is now confined within the inner volume V_1 by means of pistons, these latter can be gradually withdrawn, so that the gas slowly expands until it fills the entire volume V_2. The process can be carried out slowly, so that equilibrium conditions are maintained at all times. Moreover, a suitable supply of heat from an external source can be made to ensure that the expansion is isothermal. The change in S may now be calculated

$$S_1 = k \ln W_1.$$

$$S_2 = k \ln W_2.$$

$$S_2 - S_1 = k \ln \frac{W_2}{W_1}$$

$$= k \ln \left(\frac{V_2}{V_1} \right)^N$$

$$= Nk \ln \frac{V_2}{V_1}$$

$$= R \ln \frac{V_2}{V_1}$$

$$= R \ln V_2 - R \ln V_1.$$

This condition can be met if S is related to the volume by an equation of the form

$$S = S_0 + R \ln V,$$

where S_0 is a constant.

From this it follows that

$$dS = \frac{R}{V} . dV.$$

For an ideal gas, however,

$$PV = RT.$$

Hence,

$$dS = \frac{P . dV}{T} \qquad (4.2)$$

$$= \frac{dw}{T}$$

$$= \frac{dq}{T},$$

since, in an isothermal change, the heat absorbed, dq, must be equal to the work done by the gas, dw.

It seems then that the change in S consequent upon the isothermal expansion of an ideal gas is readily susceptible to experimental measurement, but an important limitation of our argument must be borne in mind. It was assumed that the expansion was achieved by the gradual withdrawal of pistons in such a way that equilibrium was maintained at all times. Another way of regarding this is as a process that can, at any stage, be reversed – in short, as a reversible change.

Suppose, however, that the expansion had been occasioned not by the slow withdrawal of pistons, but by the opening of a valve connecting the two chambers. In this way, the gas would have expanded into a vacuum. For the molecules of gas spontaneously to crowd back into the inner chamber is not impossible, but the statistical probability of such an event is negligibly small. Accordingly, the change is described as irreversible or spontaneous.

Even in the case of an irreversible expansion, the mathematical reasoning given above remains true as far as equation (4.2). Beyond this point, however, some modification is required. When a gas expands into a vacuum, there is at first no resisting pressure and, correspondingly, no work is done. As some of the gas enters the larger chamber, it exerts some pressure, against which the remainder of the gas will perform work, but, at the beginning of the process, it is true to say that

$$dw = 0.$$

Thus,

$$dS > \frac{dw}{T}.$$

If only a small amount of the gas is allowed to expand into a vacuum, then no heat will be required to maintain a constant temperature, since no external work is done. Hence, for a small increase in disorder,

$$dS > \frac{dq}{T}.$$

These facts can be summarised by the statement

$$dS \geq \frac{dq}{T},$$

where the equality holds for a reversible change, and the inequality for an irreversible one.

This important relationship, fundamental to all thermodynamic reasoning, has thus been established for ideal gases, but it now remains to be seen whether its validity extends to a wider context. In the foregoing arguments, it was postulated that the ideal gas could call upon some external agency that could provide or absorb heat in such a way as to maintain a constant temperature, i.e. that the chamber containing the gas was immersed in a thermostat. The thermostats generally used in a laboratory derive energy from an external electricity supply. In principle, however, it is possible to construct such a device without recourse to these means. Thus, if the apparatus consisted of a liquid at its freezing point in equilibrium with the solid phase, then it could serve as a source or recipient of heat at constant temperature without any interchange with its environment. The container of the ideal gas could then be inserted into this thermostat, so that the two formed a completely isolated system. From the considerations of Chapter 1, it follows that the large number of molecules contained in such a system would be most unlikely to sort themselves into a more orderly array. Accordingly, any irreversible change must be due to an increase in disorder. The same cannot, however, be true of a reversible change, since its reversal would then require the re-establishment of order. It follows, therefore, that a reversible change proceeding in either direction cannot be accompanied by any alteration in the degree of disorder.

Let us now suppose that one mole of the gas expands reversibly from volume V_1 to volume V_2. As already explained, the increase in disorder may be represented by the change in the quantity S, which is given by

$$dS = \frac{dq}{T},$$

or, in the case of a larger change,

$$\Delta S = \frac{q}{T}.$$

If the whole system, however, is to remain in a constant state of order, then the molecules comprising the medium of the thermostat must sort themselves into a more orderly arrangement, so that, for this part of the assembly, S must undergo a decrease numerically equal to the increase experienced by the gas. Accordingly, for the thermostat,

$$\Delta S = -\frac{q}{T}.$$

The Law of Conservation of Energy, however, requires that the heat yielded by the thermostat should be exactly equal to that absorbed by the gas, so that, both for the gas and for the thermostat medium, it is true to say:

Increase in S = heat absorbed/absolute temperature.

The reader may be a little puzzled by the requirement that the molecules in the thermostat should sort themselves into a more orderly arrangement. This may, however, easily be visualised if it is remembered that the medium consists of a liquid in equilibrium with its solid phase. The evolution of heat from such a system will require some of the liquid to solidify, the molecules thereby surrendering the freedom of movement conceded by the liquid state for the much greater constraints imposed by an orderly crystal lattice.

The arguments detailed above apply only to the reversible expansion of the gas, and their irreversible counterparts must now be considered.

When the thermostat transfers heat to the gas, the changes in the former are clearly independent of those occurring in the latter. Thus, the change in S for the thermostat only remains governed by the equation

$$\Delta S = \text{heat absorbed/absolute temperature}$$

$$= -\frac{q}{T}.$$

(In this particular case, the heat absorbed is $-q$, since heat is actually being evolved by the thermostat, whilst q refers to the heat absorbed by the gas.)

If this heat is invested in an irreversible expansion of the gas, then, for the gas only,

$$\Delta S > \frac{q}{T}.$$

It follows that, for the whole system,

$$\Delta S > \frac{q}{T} - \frac{q}{T},$$

so that

$$\Delta S > 0.$$

This is clearly in accordance with the principle already enunciated that an irreversible change should cause an increase in disorder – and, therefore, in S – for the entire system.

The results derived so far are summarised in Table 4.1. It must be remembered that an isolated system is one that is completely devoid of interchange with the rest of the universe and cannot receive or donate any matter or energy. The expressions for part of the system and for the whole of it are not really different in principle, since, for the whole of an isolated system, q is by definition equal to zero.

Table 4.1   Changes in the quantity S occurring in the whole of an isolated system or in any part of it

	Part of System	*Whole of System*
Irreversible change	$\Delta S > \dfrac{q}{T}$	$\Delta S > 0$
Reversible change	$\Delta S = \dfrac{q}{T}$	$\Delta S = 0$

Entropy

The quantity S is clearly of fundamental importance in the study of chemical and physical changes, and is accorded the name of entropy. Historically, however, it was the relationship of entropy to the absorption of heat that was discovered first, its connection with randomness being the product of later research. The traditional definition of entropy is, therefore, as follows:-

The increase in the entropy of a part of a system in which a reversible change is occurring is the quantity

$$\int \frac{\mathrm{d}q}{T},$$

where dq represents the heat absorbed at temperature T, the latter being measured in Kelvin.

The Second Law of Thermodynamics

Let us suppose that two objects are in contact with each other but isolated from the rest of the universe. Let one of the objects be at a temperature T_1, while the other is at a lower temperature T_2. Let a quantity of heat, dq, pass from the hotter object to the cooler, the two bodies being sufficiently massive for their temperatures not to be altered appreciably by this interchange. The following relationships may now be stated:-

Loss of entropy by the hotter object

$$= \frac{dq}{T_1}.$$

Gain of entropy by the cooler object

$$= \frac{dq}{T_2}.$$

For the whole system,

$$dS = \frac{dq}{T_2} - \frac{dq}{T_1}$$

$$= \left(\frac{1}{T_2} - \frac{1}{T_1} \right).dq.$$

It was, however, postulated that

$$T_1 > T_2.$$

Thus,

$$\left(\frac{1}{T_2} - \frac{1}{T_1}\right) > 0,$$

so that

$$dS > 0.$$

It can be seen, therefore, that the passage of heat from a hot body to a cooler one is an irreversible process, transfer in the opposite direction being impossible in an otherwise isolated system. It is, of course, true that machines can be constructed to transfer heat from cool surroundings to warmer ones. The presence of such a device, however, violates the condition of the isolated system, and any consideration of the entropy changes in, for example, a refrigerator must include those occurring in the machinery itself as well as in the external power source. When these are taken into account, an overall increase in entropy will again be seen to result.

These principles are enshrined in the Second Law of Thermo-dynamics, which may be stated as follows:-

Heat cannot spontaneously pass from a cooler object to a hotter one.

The Law was originally enunciated as a result of experimental findings, and the concept of entropy was derived by reasoning based upon it.

The question of the truth of the Second Law deserves some further comment. As will be realised from the arguments detailed above, the real reason why heat cannot spontaneously flow from a cooler object to a hotter one is that such a transfer would result in a decrease in entropy. Since

$$S = k \ln W,$$

a decrease in S would denote a decrease in W, i.e. a sorting of molecules into more orderly arrangements. As was explained in Chapter 1, such a tidying process is not impossible, but is of very low probability. Accordingly, during a very long period of time, it could happen that the Second Law might on occasion be disobeyed. For such numbers of molecules as are normally encountered in a reaction, however, this period would have to be inordinately long, and would exceed the estimated age of the universe by a factor of many millions of millions. The Second Law can thus, for all practical purposes, be regarded as inviolable.

The Third Law of Thermodynamics

The discussion of entropy so far has centred on the measurement or calculation of changes in this quantity rather than on its absolute value. The latter, therefore, may only be assigned if some particular condition is nominated as having zero entropy. This is the purpose of the Third Law of Thermodynamics, which may be stated as follows:-

The entropy of a crystalline solid at the absolute zero of temperature is zero.

This law is capable of a more fundamental interpretation, for entropy has already been shown to be a function of W, the number of possible arrangements leading to a given state. In a crystalline solid at the absolute zero of temperature, however, the molecules, atoms or ions are confined within a crystal lattice, and their only motion is due to their zero point energy, the minimum movement required to prevent violation of the Uncertainty Principle. It follows, therefore, that different arrangements of this state can be achieved only by the interchanging of identical particles. Since, however,

$$S = 0,$$
$$k \ln W = 0,$$

so that

$$W = 1.$$

Thus, the real meaning of the Third Law is that an exchange of positions by identical particles is not to be regarded as a different arrangement. It must, however, be remembered that this is not a fundamental law of nature, but rather a man-made limitation imposed upon the definition of W.

The Third Law gives rise to one most important consequence. In view of the more precise definition of W that it affords, it follows that a system in any given state can have only one unique value of this function, and hence of S. Entropy is, therefore, seen to be a condition of state dependent on the current condition of a system rather than on its past history, and any sequence of changes whatsoever leading to the same conditions of temperature, pressure, volume, chemical composition and phase will establish the same numerical value of the entropy.

Entropy Changes in an Ideal Gas

Reversible changes in the temperature, pressure and volume of an ideal gas are naturally associated with changes of entropy, and the magnitude of these varies according to whether the change is at constant pressure, at constant volume, isothermal or adiabatic. If one makes use of the expression

$$\mathrm{d}S = \frac{\mathrm{d}q}{T},$$

then each of these changes can readily be calculated by integration. A logical objection to this line of argument could, however, be advanced, for the considerations that led to this connection between $\mathrm{d}S$ and $\mathrm{d}q$ were developed above for isothermal changes. In all other cases, the temperature varies, so that the expression cannot be used without proof.

The case of a reversible isothermal change for one mole was discussed at the beginning of this chapter, where it was shown that

$$\Delta S = R \ln \frac{V_2}{V_1}.$$

In accordance with Boyle's Law, this can also be written

$$\Delta S = R \ln \frac{P_1}{P_2}.$$

An adiabatic change can easily be dealt with, for any adiabatic system neither receives energy from its surroundings nor donates it to them. If it is equally closed with regard to the transfer of molecules, then it must be regarded as an isolated system. Accordingly, for any reversible adiabatic change in an ideal gas,

$$\Delta W = 0,$$

so that

$$\Delta S = 0.$$

It is for this reason that adiabatic changes are sometimes called isentropic changes[1].

It was explained above that entropy is a condition of state, so that the change in entropy during any process depends only on the initial and final conditions, not on the route by which one was converted to the other. This principle enables us to calculate the entropy change for a process occurring at constant pressure. Let us consider the fate of the pressure, volume and temperature to be as follows:

$$
\begin{matrix}
P & & P \\
V_1 & \underrightarrow{\text{change at constant pressure}} & V_2 \\
T_1 & & T_2
\end{matrix}
$$

The same final result could be attained by an isothermal change followed by an adiabatic one:

$$
\begin{matrix}
P & & P' & & P \\
V_1 & \underrightarrow{\text{isothermal change}} & V' & \underrightarrow{\text{adiabatic change}} & V_2 \\
T_1 & & T_1 & & T_2
\end{matrix}
$$

In both cases, the total entropy change must be the same, and may be calculated as follows:-

---

[1] The reader may be puzzled by the fact that a gas can expand without change of entropy. According to what was said previously, an increase in volume should give the molecules more possibilities of disposition in space, so that an increase in W, and hence in S, is to be expected. An adiabatic expansion, however, is accompanied by a decrease in temperature. This reduces the velocities, and hence the freedom, of molecular motion. The one effect exactly compensates for the other, so that the entropy remains constant.

For the isothermal change,

$$\Delta S = R\ln\frac{V'}{V_1}.$$

For the adiabatic change,

$$\Delta S = 0.$$

Thus, the total entropy change in a process at constant pressure is given by

$$\Delta S = R\ln\frac{V'}{V_1}$$

$$= R\ln\left(\frac{V_2}{V_1}\cdot\frac{V'}{V_2}\right).$$

According to Charles's Law, at constant pressure,

$$\frac{V_1}{T_1} = \frac{V_2}{T_2}.$$

Therefore,

$$\frac{V_2}{V_1} = \frac{T_2}{T_1}.$$

The equations for an adiabatic change require that

$$T_1 V'^{(\gamma-1)} = T_2 V_2^{(\gamma-1)}.$$

Therefore,

$$\frac{V'}{V_2} = \left(\frac{T_2}{T_1}\right)^{\left(\frac{1}{\gamma-1}\right)}.$$

Thus,

$$\frac{V_2}{V_1}\cdot\frac{V'}{V_2}=\left(\frac{T_2}{T_1}\right)^{\left(1+\frac{1}{\gamma-1}\right)}$$

$$=\left(\frac{T_2}{T_1}\right)^{\left(\frac{\gamma}{\gamma-1}\right)},$$

and

$$\Delta S = R\ln\left(\frac{T_2}{T_1}\right)^{\left(\frac{\gamma}{\gamma-1}\right)}.$$

Now

$$\gamma = \frac{C_P}{C_V},$$

so that

$$\frac{\gamma}{\gamma-1} = \frac{C_P}{C_P-C_V}$$

$$= \frac{C_P}{R}.$$

Thus,

$$\Delta S = C_P \ln\frac{T_2}{T_1}.$$

In accordance with Charles's Law, this could also be written

$$\Delta S = C_P \ln\frac{V_2}{V_1}.$$

If the former equation is adopted, then the following interesting fact emerges.

$$\Delta S = C_P \ln \frac{T_2}{T_1}$$

$$= C_P \left(\ln T_2 - \ln T_1 \right)$$

$$= \int_{T=T_1}^{T=T_2} C_P . d(\ln T)$$

$$= \int_{T_1}^{T_2} \frac{C_P . dT}{T}$$

$$= \int_{T=T_1}^{T=T_2} \frac{dq}{T}.$$

The relationship between entropy increase and the absorption of heat is thus exactly the same as in the isothermal case.

Precisely similar arguments can be adopted for changes at constant volume. Here again, the process can be split into an isothermal component and an adiabatic one.

$$\begin{array}{ccc} P_1 & & P_2 \\ V & \xrightarrow{\text{change at constant volume}} & V \\ T_1 & & T_2 \end{array}$$

This is equivalent to

$$\begin{array}{ccccc} P_1 & & P' & & P_2 \\ V & \xrightarrow{\text{isothermal change}} & V' & \xrightarrow{\text{adiabatic change}} & V \\ T_1 & & T_1 & & T_2 \end{array}$$

For the isothermal change,

$$\Delta S = R \ln \frac{V'}{V}.$$

For the adiabatic change,

$$\Delta S = 0.$$

The total change at constant volume is, therefore, represented by

$$\Delta S = R \ln \frac{V'}{V}.$$

From the rules for isothermal changes,

$$\frac{V'}{V} = \frac{P_1}{P'}$$

$$= \frac{P_1}{P_2} \cdot \frac{P_2}{P'}.$$

For a change at constant volume,

$$\frac{P_1}{P_2} = \frac{T_1}{T_2}$$

$$= \left(\frac{T_2}{T_1} \right)^{-1},$$

while, for an adiabatic process,

$$\frac{T_2^{\gamma}}{P_2^{(\gamma-1)}} = \frac{T_1^{\gamma}}{P'^{(\gamma-1)}}.$$

Thus,

$$\frac{P_2}{P'} = \left(\frac{T_2}{T_1} \right)^{\left(\frac{\gamma}{\gamma-1} \right)}.$$

It then follows that

$$\frac{V'}{V} = \left(\frac{T_2}{T_1}\right)^{\left(\frac{\gamma}{\gamma-1}-1\right)}$$

$$= \left(\frac{T_2}{T_1}\right)^{\left(\frac{1}{\gamma-1}\right)}$$

and that

$$\Delta S = R \ln\left(\frac{T_2}{T_1}\right)^{\left(\frac{1}{\gamma-1}\right)}$$

$$= \frac{R}{\gamma-1} \ln\frac{T_2}{T_1}.$$

Now

$$\frac{R}{\gamma-1} = \frac{C_V R}{(C_P - C_V)}$$

$$= C_V.$$

Therefore,

$$\Delta S = C_V \ln\frac{T_2}{T_1}.$$

Since the change is at constant volume, an alternative formulation is

$$\Delta S = C_V \ln\frac{P_2}{P_1}.$$

Once again, a connection between increase in entropy and absorption of heat may be established.

$$\Delta S = C_V \ln \frac{T_2}{T_1}$$

$$= C_V \left(\ln T_2 - \ln T_1 \right)$$

$$= \int_{T=T_1}^{T=T_2} C_V \cdot \mathrm{d}(\ln T)$$

$$= \int_{T_1}^{T_2} \frac{C_V \cdot \mathrm{d}T}{T}$$

$$= \int_{T=T_1}^{T=T_2} \frac{\mathrm{d}q}{T}.$$

Table 4.2 Reversible changes occurring in an ideal gas

Change at constant pressure	$\Delta S = C_P \ln \dfrac{T_2}{T_1} = C_P \ln \dfrac{V_2}{V_1}$
Change at constant volume	$\Delta S = C_V \ln \dfrac{T_2}{T_1} = C_V \ln \dfrac{P_2}{P_1}$
Isothermal change	$\Delta S = R \ln \dfrac{V_2}{V_1} = R \ln \dfrac{P_1}{P_2}$
Adiabatic Change	$\Delta S = 0$

The entropy changes occurring in an ideal gas are summarised in Table 4.2. An important consequence of these relationships is that the equation

45

$$\Delta S = \int_{T=T_1}^{T=T_2} \frac{dq}{T}$$

is of completely general application for reversible processes. For irreversible changes, on the other hand, some expansion normally occurs without having to perform work against an opposing pressure or, alternatively, heat flows from a hot body to a cooler one. The latter of these cases has already been discussed, while, for the former, the arguments expounded for isothermal changes can be extended to different circumstances. Thus, for example, any irreversible expansion or compression may be itemised into an irreversible isothermal change, followed by a reversible adiabatic one. Since, for the latter, both ΔS and q are zero, the total values for these quantities will be those for the isothermal stage, so that

$$\Delta S > \int_{T=T_1}^{T=T_2} \frac{dq}{T}$$

In general, it may be said that

$$\Delta S \geq \int_{T=T_1}^{T=T_2} \frac{dq}{T},$$

where the equality holds for reversible changes, and the inequality for irreversible ones. This relationship is true for all chemical and physical changes.

Since the entropy of an isolated system can thus only increase or remain constant, and since the whole universe must be regarded as an isolated system, it follows that every irreversible change that occurs increases the entropy of the universe, while no compensating decrease can ever occur. Accordingly, the entropy of the universe must be constantly increasing. When entropy was first discovered through a consideration of energy changes in heat engines (see Chapter 9), this apparent creation of entropy out of nothing was a very puzzling concept. Only the elucidation of the connection between entropy and disorder served to explain the phenomenon.

5

Ideal Solutions

Colligative Properties

An ideal gas has already been defined as one in which the molecules are so far apart that their volume constitutes only a small part of the total space occupied and that the intermolecular forces are reduced to negligible magnitude. An analogous situation can obtain in a dilute solution, in which the solute molecules are also separated by relatively large distances. The fact that the space between them is occupied by molecules of solvent instead of being empty does little to destroy the analogy.

In the case of a non-electrolyte, the only forces to be considered are the van der Waals forces of attraction and the steric repulsion. These are weak and decline rapidly with increasing distance, so that a moderate separation of the molecules serves to promote ideal behaviour. In electrolytes, however, the electrostatic forces of attraction and repulsion are much stronger and less affected by distance. In consequence of this, an electrolyte solution requires to be more dilute in order to approximate to ideal behaviour.

From these arguments, it may be concluded that the properties of ideal solutions show certain analogies to those of ideal gases. Since the latter are dependent on the number of molecules present rather than on their chemical nature, the same must be true of ideal solutions with the proviso that the determining factor consists of the number of dissolved particles rather than that of molecular units. The distinction is important in the case of electrolytes, since each ion then functions as a separate particle.

Those properties of ideal solutions that depend only on the number of dissolved particles are said to be colligative. Before they can be discussed, however, another topic must be raised, namely that of standard states.

47

Standard States

In Chapter 4, it was shown that the entropy of one mole of an ideal gas at a given temperature can be written as

$$S = S_0 + R \ln V,$$

where S_0 is a constant.

Since

$$PV = RT,$$

this statement can be modified as follows:

$$S = S_0 + R \ln \frac{RT}{P}$$

$$= S_0 + R \ln RT - R \ln P.$$

At any given temperature, $R \ln RT$ is constant, and can be incorporated in the other constant S_0, so that

$$S = \text{constant} - R \ln P.$$

The constant in this equation is equal to the value of S when

$$P = 1,$$

and this condition is accordingly said to be the standard state of an ideal gas.

An obvious objection to this line of argument is its lack of any reference to units. The omission can be repaired by a realisation of the need for consistency among all the quantities appearing in any equation. Thus, if S is measured in joules mol^{-1} K^{-1}, the same must be true of R. Accordingly, in the equation

$$PV = RT,$$

the quantity PV must be in joules mol^{-1}. If V is in m^3 mol^{-1}, P must be in N m^{-2}. Similar arguments can be made for other systems of units.

In the case of the solute in an ideal solution, the same reasoning holds true, except that P must now represent the osmotic pressure. Thus, the entropy of one mole of the solute at a given temperature may be written either as

$$S = \text{constant} - R \ln P \text{ (where } P = \text{osmotic pressure)}$$

or as

$$S = \text{constant} + R \ln V,$$

where V is the volume containing one mole of solute.

The latter equation can be modified in several ways. In binary systems of this sort, a subscript 1 is traditionally used for quantities related to the solvent, and a subscript 2 for those concerned with the solute. Thus, let us suppose that n_1 moles of solvent contain n_2 moles of solute. Then the fraction of molecules present that are of the solvent is referred to as the molar fraction of the solvent and is denoted by N_1, the corresponding quantity for the solute possessing the symbol N_2. The following relationships should then be obvious:-

$$N_1 = \frac{n_1}{n_1 + n_2}.$$

$$N_2 = \frac{n_2}{n_1 + n_2}.$$

$$N_1 + N_2 = 1.$$

Now it is obvious that the volume containing one mole of solute is inversely proportional to the concentration of the latter and hence, in a dilute solution, to N_2. Thus,

$$V \propto \frac{1}{N_2}$$

and

49

$$S_2 = \text{constant} + R \ln V,$$

so that

$$S_2 = \text{constant} - R \ln N_2.$$

(Note that the two constants are different in value.)

A true description of a binary solution is that it is an homogeneous mixture of two substances. By convention, the component present in greater quantity is called the solvent and the other the solute, but the distinction is arbitrary and the entropy of one mole of solvent is accordingly given by

$$S_1 = \text{constant} - R \ln N_1.$$

The standard state of a liquid at a given temperature is that in which the entropy of one mole is equal to the constant in the equation above. This will be true if

$$N_1 = 1,$$

i.e. if the liquid is pure.

Similarly, the standard state for a solid is also taken as that of the pure substance.

A number of conclusions may now be drawn concerning the properties of pure liquids and ideal solutions.

Henry's Law

Suppose that a volatile solute be dissolved in a solvent of negligible volatility. The kinetic energy possessed by the solute molecules will enable a few of them to escape from the surface and form a gaseous phase above the liquid. In this vapour, however, the molecules will also be mobile, and a few of them will, on collision with the surface of the liquid, once again become dissolved in the latter. Eventually, an equilibrium will be reached, in which the number of molecules leaving the liquid surface will be exactly equal to the number entering it.

Now the entropy of one mole of the solute in the dissolved phase at a given temperature is equal to

$$S_{2 \text{ solution}} = \text{constant} - R \ln N_2,$$

while the entropy of one mole of the same substance in the gaseous phase is

$$S_{2 \text{ gas}} = \text{constant} - R \ln P_2,$$

where P_2 is the partial pressure of solute vapour at equilibrium. It follows then that, if dn mole of solute passes from the liquid to the vapour phase, the latter will gain in entropy by a quantity

$$(\text{constant} - R \ln P_2).dn,$$

while the former will lose by

$$(\text{constant} - R \ln N_2).dn.$$

Since the two constants can be combined into one, the total increase in entropy will be

$$(\text{constant} - R \ln P_2 + R \ln N_2).dn$$

$$= \left(\text{constant} - R \ln \frac{P_2}{N_2}\right).dn.$$

If the change occurs at equilibrium, however, it must be reversible, so that the total increase in entropy must be zero. Accordingly,

$$\left(\text{constant} - R \ln \frac{P_2}{N_2}\right).dn = 0.$$

$$R \ln \frac{P_2}{N_2} = \text{constant}.$$

This means that the ratio P_2/N_2 must be constant, so that

$$P_2 \propto N_2.$$

If, however, c_2 is the concentration of the solute, then, to a good approximation,

$$N_2 \propto c_2.$$

Therefore,

$$P_2 \propto c_2$$

Thus, the equilibrium partial pressure of the solute vapour is proportional to the concentration of the same substance in the solution. This statement is known as Henry's Law.

The Lowering of Vapour Pressure

The discussion of Henry's Law was concerned with the case in which a volatile solute was dissolved in a solvent of much lower volatility. As was previously pointed out, however, the distinction between solute and solvent is dictated by convention rather than by any chemical reality. All the reasoning of the previous section can, therefore, be applied to a volatile solvent containing a non-volatile solute. In this case, the conclusion will be that the vapour pressure of the solvent is proportional to its molar fraction in the solution, i.e. that

$$P_1 \propto N_1.$$

This reasoning can be extended further in the following way:-

$$P_1 = \text{constant} \times N_1.$$

If the solvent is taken in the pure state and its vapour pressure in this condition is P_1^*, then

$$N_1 = 1,$$

so that

$$P_1^* = \text{constant.}$$

Thus, the constant of proportionality between P_1 and N_1 is equal to the vapour pressure of the pure solvent, so that

$$P_1 = P_1^* N_1, \qquad (5.1)$$

and

$$N_1 = \frac{P_1}{P_1^*}.$$

It has already been shown that

$$N_1 + N_2 = 1.$$

so that

$$\frac{P_1}{P_1^*} = N_1$$

$$= 1 - N_2,$$

and

$$N_2 = 1 - \frac{P_1}{P_1^*}$$

$$= \frac{P_1^* - P_1}{P_1^*} \qquad (5.2)$$

From equation (5.1), it can readily be seen that the vapour pressure of a solution must be less than that of the pure solvent. It is equation (5.2), however, that gives the quantitative relationship in its most useful form. Here, P_1^* is the vapour pressure of the pure solvent, and P_1 that of the solvent in a solution. The quantity $(P_1^* - P_1)$ is known as the lowering of vapour pressure, while

$$\frac{P_1^* - P_1}{P_1^*}$$

is the relative lowering of vapour pressure. As can be seen from equation (5.2), the relative lowering of vapour pressure of the solvent is equal to the molar fraction of the solute. This statement is known as Raoult's Law.

The Elevation of the Boiling Point

The previous section proved that the vapour pressure of a volatile solvent was reduced in the presence of a solute. A later section will show that the vapour pressure of a liquid is raised by an increase in temperature. It is clear, therefore, that a solution will have to be heated to a higher temperature if the vapour pressure of the solvent is to become equal to the atmospheric pressure, so that boiling can occur.

At first sight, the entropy considerations concerned in the lowering of vapour pressure and the elevation of the boiling point may seem to be similar, since both cases are involved in the transfer of molecules between the liquid and vapour states. In actual fact, however, the two cases are very different. In the discussion of the lowering of vapour pressure, a very small quantity of solvent was allowed to evaporate at the same time as an equal quantity condensed. The total change in energy was, therefore, zero, and the system could be regarded as isolated, so that, at equilibrium,

$$\Delta S = 0.$$

In contrast, a boiling liquid is one in which a large quantity of substance enters the vapour phase without any corresponding condensation. This can only be achieved if heat is constantly being supplied from an external source, so that the system cannot be regarded as isolated. The change in entropy at equilibrium is, therefore, given by the expression

$$\frac{\text{heat absorbed}}{\text{absolute temperature}}.$$

The heat required for the evaporation of one mole of the solvent is known as the molar latent heat of vapourisation, and will be denoted by the symbol L. If the boiling point of the pure solvent is T and the elevation of the boiling point ΔT, then, for one mole,

$$\Delta S = \frac{L}{T + \Delta T}.$$

If the atmospheric pressure is denoted by P_a, then the entropy of one mole of solvent in the vapour phase is

$$\text{constant} - R \ln P_a.$$

Its entropy in the liquid phase is

$$\text{constant} - R \ln N_1.$$

The increase in entropy on evaporation is given by the difference between these two expressions, but the two constants may be combined into one, and the term $R \ln P_a$ may be incorporated in this. It therefore follows that

$$\begin{aligned}
&\text{entropy of vapourisation of one mole} \\
&= (\text{constant} - R \ln P_a) - (\text{constant} - R \ln N_1) \\
&= \Delta S_0 + R \ln N_1 \text{ (where } \Delta S_0 \text{ is a constant)}
\end{aligned}$$

$$= \frac{L}{T + \Delta T} \qquad (5.3)$$

If the pure solvent were taken instead of a solution, then equation (5.3) would have to be modified by the considerations that

$$N_1 = 1$$

and

$$\Delta T = 0.$$

The resulting equation would be

$$\Delta S_0 = \frac{L}{T} \qquad (5.4)$$

The substitution of this into equation (5.3) leads to the following result:

$$\frac{L}{T} + R \ln N_1 = \frac{L}{T + \Delta T}.$$

$$R \ln N_1 = \frac{L}{T + \Delta T} - \frac{L}{T}.$$

$$R\ln(1 - N_2) = -\frac{L.\Delta T}{T(T + \Delta T)} \qquad (5.5)$$

According to Maclaurin's Theorem,

$$\ln(1 + x) = x - x^2/2 + x^3/3 - x^4/4 + \ldots$$

If x is small, the terms on the right-hand side of this equation will become progressively smaller, and, if x is small enough, it may be said that

$$\ln(1 + x) \approx x.$$

In a dilute solution, N_2 will be small, so that the left-hand side of equation (5.5) may be approximated as follows:

$$R\ln(1 - N_2) \approx -RN_2.$$

The small value of ΔT also affords a means of approximation for the right-hand side of equation (5.5).

$$-\frac{L.\Delta T}{T(T + \Delta T)} \approx -\frac{L.\Delta T}{T^2}.$$

The equation may now be rewritten in the following form:

$$- RN_2 = -\frac{L.\Delta T}{T^2}.$$

Therefore,

$$\Delta T = \left(\frac{RT^2}{L}\right) N_2.$$

Thus, the elevation of the boiling point is equal to the molar fraction of the solute multiplied by the quantity

$$\frac{RT^2}{L},$$

which is known as the ebullioscopic constant of the solvent.

The Depression of the Freezing Point

The boiling of a solution involves the transfer of solvent from the liquid phase, in which it is in admixture with the solute, to the vapour phase, in which it is pure. The freezing of a solution similarly involves transfer of solvent from a mixture into a separate phase in which it is pure. The two processes are, therefore, exactly analogous except that, in the former, pure solvent represents the state of higher energy, while, in the latter, it is of lower energy. Apart from this point of principle and one or two details, however, the mathematical treatments of the two have a close resemblance, as may be seen below:-

Entropy of one mole of solvent in solution
$$= \text{constant} - R \ln N_1.$$
Entropy of one mole of solvent in the solid phase $= \text{constant}.$

The increase in entropy on the melting of one mole of solid is given by the subtraction of the second of these equations from the first.

Increase in entropy on the melting of one mole of solid
$$= \Delta S = \Delta S_0 - R \ln N_1,$$

where ΔS_0 = constant.

This increase in entropy may, however, also be written as

$$\Delta S = \frac{L}{T + \Delta T},$$

where L = molar latent heat of fusion,

T = freezing point of the pure solvent,

and ΔT = freezing point of solution – freezing point of pure solvent.

The two expressions for ΔS may be equated.

$$\Delta S_0 - R \ln N_1 = \frac{L}{T + \Delta T} \qquad (5.6)$$

In the case of the pure solvent,

$$N_1 = 1 \text{ and } \Delta T = 0.$$

Thus,

$$\Delta S_0 = \frac{L}{T}.$$

Substitution of this into equation (5.6) yields the following:

$$\frac{L}{T} - R \ln N_1 = \frac{L}{T + \Delta T}.$$

Hence,

$$R \ln N_1 = \frac{L}{T} - \frac{L}{T + \Delta T}.$$

As already explained,

$$R \ln(1 - N_2) = \frac{L.\Delta T}{T(T + \Delta T)}.$$

$$\ln(1 - N_2) \approx -N_2$$

and

$$T(T + \Delta T) \approx T^2.$$

Thus,

$$- RN_2 = \frac{L.\Delta T}{T^2}$$

and

$$\Delta T = -\left(\frac{RT^2}{L}\right)N_2.$$

It will be seen that, this time, ΔT is a negative quantity, the fundamental reason for this being in the original subtraction of entropies, which was carried out in the opposite sense to that of the previous section.

The quantity RT^2/L is known as the cryoscopic constant of the solvent.

The Increase of Vapour Pressure with Temperature

Let a liquid be in equilibrium with its vapour at a temperature T_1, the vapour pressure being P_1. Let the molar latent heat of vapourisation be L. The following relationships may then be written.

Entropy of dn mole of substance in vapour form
= (constant − R ln P_1).dn.
Entropy of dn mole of substance in liquid form
= constant × dn.

If the difference between the two constants is written as ΔS_0, then the application of a little heat from an external source will cause dn mole to evaporate, so that

entropy of vapourisation of dn mole = (ΔS_0 − R ln P_1).dn

$$= \frac{L.\mathrm{d}n}{T_1}.$$

Hence,

$$\Delta S_0 - R\ln P_1 = \frac{L}{T_1} \qquad (5.7)$$

Similar reasoning for a temperature T_2 at which the vapour pressure is P_2 leads to the result

$$\Delta S_0 - R\ln P_2 = \frac{L}{T_2} \qquad (5.8)$$

Both ΔS_0 and L are somewhat temperature dependent, but they may be regarded as approximately constant if T_1 and T_2 are not too far apart. The subtraction of equation (5.8) from equation (5.7) with subsequent division by R then yields the relationship between the vapour pressure of a pure liquid and the temperature:

$$\ln\frac{P_2}{P_1} = \frac{L}{R}\left(\frac{1}{T_1} - \frac{1}{T_2}\right).$$

As may be seen from this, an increase in temperature leads to an increase in vapour pressure.

The result obtained may also be derived from the Clapeyron–Clausius equation, which will be discussed later.

Trouton's Constant

If the molecules of a liquid are attracted to one another only by weak van der Waals forces, then the number of possible arrangements should not vary greatly between one substance and another. The same may be said of the molecules in the vapour phase at atmospheric pressure. The degree of randomisation introduced by the transfer of one mole of substance from the liquid phase to the vapour at atmospheric pressure, i.e. by boiling, should, therefore, ideally be the same for all substances. Another way of saying this is that the entropy of boiling should be equal for all liquids. Since the heat absorbed when one mole of liquid boils is equal to the molar latent heat of vapourisation, it follows that the expression

$$\frac{\text{molar latent heat of vapourisation}}{\text{boiling point}}$$

should ideally have the same value for all liquids. This is known as Trouton's constant, and its value is 88 J mol^{-1} K^{-1}.

Of course, this discussion of entropy of vapourisation is somewhat oversimplified, for the degree of randomness in either the liquid or the vapour phase can scarcely be identical in substances with greatly differing boiling points. Although various attempts have been made to replace Trouton's constant by quantities measured under more rigorously specified conditions, the original concept nevertheless remains a useful one.

A more serious objection to the treatment above stems from the assumption that the molecules in the liquid are subject only to weak van der Waals forces. Clearly, the presence of strong dipole moments will invalidate this stipulation, while the occurrence of hydrogen bonds will be of even greater significance. One would, therefore, expect a strongly dipolar or hydrogen bonded substance to display an abnormal Trouton constant, an anticipation that is adequately borne out in practice.

Chapters 4 and 5 are concerned with substances in the ideal state. This condition can often be approximately realised in practice, and forms a useful starting point for much thermodynamic reasoning. It frequently happens, however, that substances must be used under conditions considerably removed from the ideal state. In such a case, they are referred to as real gases or real solutions, and it is these which must engage our attention in Chapter 6.

6

Deviations from Ideal Behaviour

As already explained, an ideal gas is one in which the molecules are so far apart that the volume which they occupy and the forces acting between them can be regarded as negligible. An ideal solution is one in which the same can be said of the molecules or ions of the solute. The laws governing the behaviour of ideal gases assume a particularly simple form embodied in the equations

$$PV = RT \qquad (6.1)$$

and

$$C_P - C_V = R, \qquad (6.2)$$

where P = pressure,
V = volume occupied by one mole,
R = gas constant,
T = absolute temperature,
C_P = molar thermal capacity at constant pressure,
and C_V = molar thermal capacity at constant volume.

Ideal solutions obey an equation of the same form as (6.1), except that P represents the osmotic pressure, while V denotes the volume containing one mole of solute.

In practice, however, gases and solutions show greater or lesser degrees of deviation from these simple laws. This is particularly true of gases under high pressure or at low temperature and of solutions that are not very dilute. These materials are referred to as real gases and real solutions. Any discussion of real gases should, however, be prefaced by a further consideration of the isothermal expansion of an ideal gas.

In Chapter 3, it was explained that, when an ideal gas expands isothermally from a pressure P_1 to a lower pressure P_2, the work done by one mole of the gas (w) is given by the relationships

$$w = \int_{V_1}^{V_2} P.dV = RT \ln \frac{P_1}{P_2}, \qquad (6.3)$$

where V_1 and V_2 are the initial and final volumes of one mole. There is, however, another way of expressing this quantity.

In accordance with equation (6.1),

$$PV = RT.$$

If T = constant, then PV = constant.

By differentiation,

$$P.dV + V.dP = 0.$$

Thus,

$$P.dV = -V.dP.$$

$$w = \int_{V=V_1}^{V=V_2} -V.dP$$

$$= \int_{P_1}^{P_2} -V.dP \qquad (6.4)$$

Substitution from equation (6.1) will lead to the same result as was obtained in equation (6.3), but equation (6.4) is important in its own right.

Real Gases

In real gases, equation (6.1) is at best only an approximation. In some cases, its use does not occasion serious errors, but in others the deviation is considerable. Numerous attempts have been made to express the behaviour of a real gas more accurately, the best known of them being the van der Waals equation:

$$\left(P+\frac{a}{V^2}\right)(V-b)=RT,$$

where a and b are constants.

The science of thermodynamics, however, is less concerned with the equations of state for a real gas than with energy relationships. These are best portrayed by the introduction of some new quantities.

The first of these is the fugacity, usually denoted by $P*$ with any appropriate subscript. It is somewhat analogous to pressure, and is measured in the same units. The work done by one mole of a real gas expanding isothermally from fugacity P_1^* to fugacity P_2^* is given by

$$w=RT\ln\frac{P_1^*}{P_2^*} \qquad (6.5)$$

Since this work is a quantity that can be measured, equation (6.5) affords a means of defining the ratio of two fugacities. The definition of the absolute value of fugacity, however, necessitates a further stipulation, namely that, at very low pressures, the fugacity and pressure approach each other in value, and, in the limiting case of infinitesimally small pressure, they become numerically equal.

In this way, the exact value of a fugacity can be established, but the relationship between this quantity and the pressure is capable of a different interpretation. Suppose that one mole of a real gas at temperature T be made to occupy a volume V. For this purpose, a pressure P will be required. If the gas were ideal, then the pressure needed would be slightly different. This latter quantity is known as the ideal pressure, and will be denoted by P_i.

Let the quantity α be taken to represent the difference in volume occupied by one mole of a real gas and that which would contain an equal amount of an ideal gas at the same temperature and pressure. Thus,

$$\alpha=V-\frac{RT}{P} \qquad (6.6)$$

For a large range of conditions, α may be taken as more or less constant.

Now, by equation (6.4), the isothermal compression of the gas from pressure P_1 to pressure P involves the performance by the gas of work w, where

$$w = \int_{P_1}^{P} -V.dP.$$

The quantity w will, of course, be negative. Substitution from equation (6.6) gives

$$w = \int_{P_1}^{P} \left(-\alpha - \frac{RT}{P} \right).dP$$

$$= \alpha(P_1 - P) + RT \ln \frac{P_1}{P}.$$

By equation (6.5),

$$w = RT \ln \frac{P_1^*}{P^*},$$

where P_1^* and P^* are the initial and final fugacities. Thus,

$$RT \ln \frac{P_1^*}{P^*} = \alpha(P_1 - P) + RT \ln \frac{P_1}{P}.$$

In the limit as P_1 tends towards zero, P_1^* tends towards P_1, so that

$$RT \ln \frac{P_1}{P^*} \approx -\alpha P + RT \ln \frac{P_1}{P}.$$

$$RT \ln \frac{P}{P^*} = -\alpha P.$$

$$RT \ln \frac{P^*}{P} = \alpha P$$

$$= PV - RT$$

$$= \frac{PRT}{P_i} - RT.$$

$$\ln\frac{P^*}{P} = \frac{P}{P_i} - 1.$$

The ratio P^*/P will not usually differ very greatly from 1, and, as was explained in Chapter 5, Maclaurin's Theorem thus provides the approximation

$$\ln\frac{P^*}{P} \approx \frac{P^*}{P} - 1.$$

Hence,

$$\frac{P^*}{P} - 1 \approx \frac{P}{P_i} - 1.$$

$$P^2 = P_i P^*.$$

This last equation can be expressed in words by the statement that the pressure acting on a gas is approximately equal to the geometric mean of the ideal pressure and the fugacity.

Real Solutions

In Chapter 5, it was shown that, if a solute in an ideal solution has a molar fraction of N_2, then its entropy, S_2, is given by the equation

$$S_2 = \text{constant} - R \ln N_2. \qquad (6.7)$$

If the solution contains n_1 molecules of solvent and n_2 molecules of solute, then

$$N_2 = \frac{n_2}{n_1 + n_2}.$$

In general, n_1 is very much greater than n_2, so that slight variations of the latter have little effect on the sum $(n_1 + n_2)$. Accordingly,

$$n_1 + n_2 = \text{constant.}$$

This last condition, however, is true only if the total volume of the solution is kept constant. In general,

$$n_1 + n_2 \propto volume.$$

Therefore,

$$N_2 \propto \frac{n_2}{volume}.$$

$$N_2 \propto c_2,$$

where c_2 is the concentration of the solute. Equation (6.7) for an ideal solution can now be rewritten as

$$S_2 = \text{constant} - R \ln c_2. \qquad (6.8)$$

In the case of a real solution, the divergence of behaviour can most conveniently be expressed by the substitution of another quantity in place of the concentration. This new quantity is known as the activity, and, if a_2 be used as its symbol for the solute, then equation (6.8) becomes

$$S_2 = \text{constant} - R \ln a_2. \qquad (6.9)$$

Once again, it can be seen that equation (6.9) by itself merely defines the ratio of two activities, for, if a_2 and a_2^* are the activities corresponding with entropies S_2 and $S_2{}^*$, then

$$S_2 = \text{constant} - R \ln a_2$$

and

$$S_2^* = \text{constant} - R \ln a_2^*,$$

so that

$$S_2 - S_2^* = R \ln \frac{a_2^*}{a_2}.$$

As has already been indicated, changes in entropy can be measured, so that the value of a_2^*/a_2 can be determined. The activity of a given state can be defined by the additional stipulation that, at infinitesimally low concentration, activity and concentration become numerically equal.

From the foregoing, it will readily be seen that activity has the same dimensions as concentration. It is accordingly measured in the same units.

Another quantity, known as the activity coefficient, is defined by the equation

activity = activity coefficient × concentration.

Clearly, activity coefficient is a dimensionless quantity and, therefore, has no units.

7

Chemical and Physical Equilibria

The discussion in Chapter 1 indicated that the science of thermo-dynamics is primarily that which identifies the driving force impel-ling chemical and physical changes to occur. Any such change progresses towards a state in which the system is stable, that is to say one in which equilibrium has been reached. A change that converts an unstable condition into a stable one can proceed in one direction only, and is, therefore, irreversible, whilst one that transforms one state of equilibrium into another can go in either direction, and is, therefore, reversible. Clearly, the study of equilibria is crucial to an understanding of chemical and physical change, and must now be discussed. The simplest equilibria are those involving alterations of state of pure substances, and these will, therefore, form the opening section of the present chapter.

Melting and Boiling Points

The actual value of the melting point or boiling point of a material arises from quite simple considerations. Let us begin with the former.

The entropy of a pure substance in the crystalline state is governed by the geometry of the crystal structure, the vibrational degrees of freedom of the molecule and similar considerations. The entropy of the same substance in the liquid phase is naturally much greater owing to the freedom of motion of the molecules and hence the additional opportunities for randomisation. The absorption of heat by either the solid or the liquid will result in an increase in entropy, but this will be small in comparison with the increase occasioned by a change from the solid to the liquid phase. Since the molecules or ions of a solid are in very close proximity to one another, inter-molecular or interionic forces are strong, and, for melting to occur, these forces must be overcome by the supply of latent heat. This latter quantity is slightly, but only slightly, temperature dependent.

Let the increase in entropy resulting from the melting of one mole be ΔS, and let the molar latent heat of fusion be L. Let the temperature be T. Both ΔS and L are somewhat dependent on the temperature, but T is obviously much more so, so that the former two quantities can be considered approximately constant. In the expression

$$\Delta S \geq \frac{L}{T},$$

therefore, ΔS and L are quantities determined by fundamental considerations of crystal structure, molecular vibrations, intermolecular or interionic forces and similar factors. Accordingly, there will be only one value of T for which

$$\Delta S = \frac{L}{T},$$

and this will depend on the nature of the substance under consideration. At this temperature, the phase change will be reversible, so that the melting point will have been attained. If the temperature is raised slightly, ΔS and L will not be greatly affected, but L/T will clearly be decreased, so that

$$\Delta S > \frac{L}{T}.$$

Hence, the melting will be an irreversible process, as should be the case at a temperature above the melting point.

Suppose that the temperature is decreased slightly below the melting point. Then

$$|\Delta S| < \frac{|L|}{T}.$$

This condition can clearly not lead to any melting, but let us consider what would happen if the liquid solidified at such a temperature. Since the conversion from the liquid phase to the solid one is accompanied by a decrease in the freedom of movement enjoyed by molecules or ions, there must be a corresponding loss of entropy. Similarly, the particles affected by intermolecular or interionic forces of attraction will move closer together, and the corresponding loss of potential energy will manifest itself in the evolution of heat. Accordingly, ΔS and L will now become negative quantities. It has, however, already been shown that the numerical value of ΔS will be less than that of L/T. Hence, at a temperature below the melting point, it is the process of solidification for which

Increase in entropy $>$ heat absorbed/temperature,

so that freezing becomes an irreversible change.

The reader may object that the argument given above involves a decrease in entropy, while all the arguments previously developed indicate that entropy should always increase or remain constant. The apparent contradiction disappears when it is recollected that the mixture of solid and liquid does not represent an isolated system. The heat evolved during the freezing process escapes to the environment, where it promotes more vigorous kinetic motion of molecules, and thus greater randomisation. If the whole of the system is considered, then an overall increase in entropy will be found to occur when a substance solidifies at a temperature below its melting point.

Exactly the same considerations apply to the boiling of a pure liquid, but here another effect comes very much into evidence. The entropy of a mole of substance in the vapour phase is very much dependent on the pressure in accordance with the equation

$$S = \text{constant} - R \ln P^*,$$

where P^* is the fugacity. Accordingly, the change in entropy on conversion of one mole of substance from liquid to vapour is also pressure dependent, and the same must be true of the boiling point.

A similar, but much smaller, effect due to pressure will also modify the exact value of the melting point of a solid.

The reader may like to compare this section with the discussion in Chapter 5 concerning the melting points, boiling points and vapour pressures of solutions.

Solubility of an Unionised Solid

The entropy of one mole of an unionised substance in dilute solution has already been shown to be of the form

$$S = \text{constant} - R \ln c,$$

where c is the concentration.

The entropy of one mole of the same substance in the solid phase may be regarded as constant, so that the change in entropy due to the entry into solution of one mole is given by

$$\Delta S = \text{constant} - R \ln c.$$

The dissolving of the substance causes the breaking up of a crystal lattice, and, as already explained, this requires the absorption of energy. The effect is partly offset by the solvation of the molecules, which causes the release of potential energy, but the net effect is still usually an intake of energy. Let us call this molar heat of solution q. If the temperature is T, then

$$\text{constant} - R \ln c \geq \frac{q}{T}.$$

Obviously, with a given value of T, there will be only one value of c for which

$$\text{constant} - R \ln c = \frac{q}{T},$$

and this value will represent the solubility of the material, since solution or crystallisation will then be a reversible process. It is equally obvious that, if T is increased, a greater value of c will be required to satisfy the equation above. Thus, the solubility of a solid is usually increased by a rise in temperature. In a few cases, the energy evolved on solvation exceeds that absorbed for the breaking of the crystal lattice. Solution then becomes an exothermic process, so that q is negative. By the reasoning given above, the solubility of such a substance decreases with a rise in temperature, and is said to be retrograde. In a very few cases, such as that of sodium chloride, the value of q is close to zero, so that the solubility is almost independent of the temperature.

Two further points deserve to be made. Firstly, the expression

$$\text{constant} - R \ln c$$

for the entropy of one mole in solution is approximate. For greater accuracy, the activity should have been used instead of the concentration. Secondly, the considerations above refer only to the solubility of solids. If the solubility of a gas is being considered, then it is quite incorrect to regard the entropy of the pure substance as constant.

Solubility of an Ionised Solid

Let us consider a substance of formula $A_m B_n$, which ionises so that

$$A_m B_n \rightarrow mA^{n+} + nB^{m-}.$$

If W_1, W_2, W_3 etc. be the respective probabilities that 1, 2, 3, ... ions of A^{n+} are contained in a volume V defined in a container of solvent, then

$$W_1 \propto V,$$

$$W_2 \propto V^2,$$

$$W_3 \propto V^3$$

and so on. In general, if N is the Avogadro number,

$$W_{mN} \propto V^{mN}.$$

Since the volume containing mN ions of the type A^{n+} is inversely proportional to the concentration, c_A, it would seem that

$$W_{mN} \propto \left(\frac{1}{c_A}\right)^{mN}.$$

Actually, the interionic forces in electrolytes are much stronger than the intermolecular ones in unionised substances, so that the approximation of using concentration instead of activity gives rise to greater inaccuracies. Accordingly, it is better to write

$$W_{mN} \propto \left(\frac{1}{a_A}\right)^{mN},$$

where a_A is the activity of the ions A^{n+}. The corresponding expression for the anions is

$$W_{nN} \propto \left(\frac{1}{a_B}\right)^{nN},$$

where a_B is the activity. The total probability for one mole is, therefore, given by W, where

$$W = W_{mN} \times W_{nN} \propto \left(\frac{1}{a_A}\right)^{mN} \times \left(\frac{1}{a_B}\right)^{nN},$$

and the corresponding entropy, S, emerges as

$$S = k \ln W$$
$$= \text{constant} - R \ln (a_A^m.a_B^n).$$

Once again, the entropy in the solid state may be regarded as constant, so that the entropy of solution of one mole can be written as

$$\text{constant} - R \ln (a_A^m.a_B^n).$$

With the same nomenclature as before,

$$\text{constant} - R \ln \left(a_A^m.a_B^n\right) \geq \frac{q}{T},$$

so that, at a given temperature, only one value of the expression $a_A^m . a_B^n$ will satisfy the conditions for equality.

The quantity $a_A^m . a_B^n$ for a saturated solution is known as the solubility product of the electrolyte $A_m B_n$. In the case of substances of very low solubility, activities can be replaced by concentrations, but otherwise the use of activities is preferable.

Distribution between Solvent Layers

Let the solubilities in moles per unit volume of an unionised substance in two solvents be c_1 and c_2, and the corresponding molar heats of solution be q_1 and q_2. Then,

$$\text{constant} - R \ln c_1 = \frac{q_1}{T},$$

and

$$\text{constant} - R \ln c_2 = \frac{q_2}{T}.$$

The constant on the left-hand side depends only on the constant of proportionality between W and V and the entropy of the solid. It must, therefore, be equal for both equations. Accordingly, by subtraction,

$$R \ln \frac{c_2}{c_1} = \frac{q_1 - q_2}{T} \qquad (7.1)$$

Consider now the case where the two solvents are immiscible, and the solute, in an unsaturated solution, distributes itself between them, so that, at equilibrium, the molar concentrations are K_1 and K_2 respectively. Let dn mole of solute transfer itself from the second solvent to the first. The gain in entropy will be

$$[\text{constant} - R \ln K_1 - (\text{constant} - R \ln K_2)].dn$$

$$= \left(R \ln \frac{K_2}{K_1} \right).dn,$$

77

and the heat absorbed will be

$$(q_1 - q_2).dn.$$

Thus,

$$\left(R \ln \frac{K_2}{K_1} \right).dn = \frac{(q_1 - q_2).dn}{T},$$

and

$$R \ln \frac{K_2}{K_1} = \frac{q_1 - q_2}{T} \qquad (7.2)$$

From equations (7.1) and (7.2), it may be seen that

$$\frac{K_2}{K_1} = \frac{c_2}{c_1},$$

so that the ratio of the concentrations is equal to that of the solubilities. This ratio is referred to as the partition coefficient.

In this calculation, two approximations have been made. Firstly, concentrations have been used instead of activities. Provided, however, that the concentrations are not too great, the error introduced in this way will not be particularly serious. Secondly, it was assumed that the molar heat of solution was independent of the concentration of the solute. This is not strictly accurate, but approximates fairly closely to the truth. It was also assumed that no chemical change, such as association or dissociation, occurs during the transfer from one solvent to the other.

By way of final comment, it may be seen that the properties of the solvents have entered these considerations only in so far as they affect the molar heat of solution. Indeed, the solvents could be eliminated from the deliberations altogether, and the partitioning be regarded as occurring between two phases. In this way, it can be seen that Henry's Law and Raoult's Law, discussed in Chapter 5, are special cases of the distributions just discussed. The further consequences of distribution between phases are discussed in the next section.

The Phase Rule

In order to discuss the distribution of substances between phases, it is necessary to define more clearly what is meant by a phase and by a component.

A phase is a physical state with distinct boundaries. Thus, a solid, a liquid and a gas are all separate phases. Any homogeneous mixture – such as a mixture of gases or a solution of a solid in a liquid, a liquid in another liquid, a gas in a liquid, a gas in a solid or a solid in another solid – constitutes a single phase, whilst liquids forming separate layers or solids in admixture but not in a solid solution are regarded as separate phases.

For a consideration of phase equilibria, it is necessary to define the number of components of a system. A component is defined as a material, or mixture of materials, having a particular percentage composition of elements. The number of components of a system is the number of components in terms of which the composition of every phase can be specified by the use of suitable coefficients, even if these are zero. Two examples will serve to clarify this point.

Consider the system

$$CaCO_3 \rightleftarrows CaO + CO_2.$$

In this reaction, there are three phases, namely solid calcium carbonate, solid calcium oxide and gaseous carbon dioxide. The compositions of all three can, however, be stated in terms of calcium oxide and carbon dioxide as follows:-

$$Calcium\ carbonate = CaO + CO_2;$$
$$Calcium\ oxide = CaO + 0\ CO_2;$$
$$Carbon\ dioxide = 0\ CaO + CO_2.$$

The system accordingly has two components, calcium oxide and carbon dioxide.

By contrast, consider the following system:

$$NH_4Cl \rightleftarrows NH_3 + HCl$$

This system has two phases, namely a solid and an homogeneous mixture of gases. Although the solid consists of ammonium chloride and the gaseous phase of ammonia and hydrogen chloride, the percentage composition of elements is the same for both phases, so that the system has only one component with percentage composition equal to that of ammonium chloride.

Note that the number of components of a system is the number in terms of which *every* phase can be described.

The complete description of any system requires more than a mere statement of the number of phases and components. Temperatures, pressures and the concentrations of the different components in the different phases must also be included. If, however, it is stipulated that the system shall be at equilibrium, then certain values of these variables will be inconsistent with this requirement, so that the specification of some necessarily implies certain magnitudes of the others. The minimum number of variables that must be stated to provide a complete description of a system at equilibrium is known as the number of degrees of freedom of the system.

According to the Second Law of Thermodynamics, a system can be at equilibrium only if its temperature is uniform throughout. For obvious reasons, the same must be true of the pressure, so that, at equilibrium, only one value for each of these two variables needs to exist.

If the system consists of P phases each having C components, then the number of concentrations would seem to be PC. Further consideration, however, shows that not all of these need to be specified. It is clear that the composition of each phase must add up to 100%, so that, if the concentrations of all but one of the components are known, the last one can be deduced. Thus, for each phase, there is one component the concentration of which need not be stated, and the number of concentrations to be defined is thus reduced to

$$PC - P.$$

Not all possible values of these will, however, be consistent with a state of equilibrium. The previous section demonstrated that, at equilibrium under given conditions of temperature (and pressure if gases are involved), the ratio of concentrations in two phases will be constant. Thus, a statement of the concentration in one phase suffices to define the concentrations of the same component in all the other phases. For each component, therefore, $(P - 1)$ concentrations need not be specified, and the total number of concentrations that must be independently given is reduced to

$$PC - P - C(P - 1).$$

A complete description of the system must include the temperature and pressure, so that the number of degrees of freedom, F, is given by

$$F = PC - P - C(P - 1) + 2$$
$$= C + 2 - P,$$

so that

$$P + F = C + 2.$$

This last equation is known as the Phase Rule, and shows how the number of phases, P, degrees of freedom, F, and components, C, are related. It is important to remember that the rule applies only to systems at equilibrium, and that the only degrees of freedom taken into consideration are the temperature, pressure and concentrations. Such variables as electrical forces, surface tension etc. are excluded.

In order to illustrate the workings of the Phase Rule, two examples will be discussed.

Consider a saturated solution of a non-volatile solute, A, in a volatile solvent, B. A saturated solution may be considered as a solution in contact with the solute in the solid phase. There are thus three phases, each of which can be defined in terms of two components as follows:-

Solid $= A + 0B$.
Liquid $= xA + yB$ (where x and y represent the proportions).
Gas $= 0A + B$.

Hence,

$$P = 3$$

and

$$C = 2.$$

According to the Phase Rule,

$$P + F = C + 2,$$

so that

$$F = 1.$$

This means that only one variable needs to be defined in order for the system to be fully described. Thus, for example, if the temperature is given, the concentration and vapour pressure may be deduced if suitable tables are to hand.

The second example consists of an unsaturated solution of A in B. Here, there is no solid phase, so that

$$P = 2$$

and

$$C = 2.$$

Accordingly,

$$F = 2.$$

Thus, the system is not fully described by the temperature alone, and both the temperature and the concentration must be stated if the vapour pressure is to be deduced.

The reader may like to apply this reasoning to the dissociation of calcium carbonate and of ammonium chloride described above. It will be found that each of them has one degree of freedom. It is interesting to compare the case of calcium carbonate with that of cupric oxide, which decomposes according to the equation

$$4CuO \rightleftarrows 2Cu_2O + O_2.$$

Here, however, the two solids form a solid solution, i.e. an homogeneous mixture constituting one phase, while the gas provides a separate phase. In consequence, there are only two phases, and hence two degrees of freedom. This means that, at a given temperature, the equilibrium pressure of oxygen depends on the relative proportions of the two oxides in the solid phase.

The terms univariant, bivariant etc. are sometimes used for systems with one, two etc. degrees of freedom. In addition, there are also invariant systems with no degrees of freedom. A good example of the latter is a pure substance melting and boiling at its triple point. The mere fact that the triple point has been reached then affords a complete description of the system without the need to state any further parameters.

Reversible Chemical Reactions

Let us consider a chemical reaction in which reagents E, F,... react reversibly to form products L, M,... in accordance with the following equation:

$$eE + fF + \ldots \rightleftarrows lL + mM + \ldots$$

In accordance with previous reasoning, let W_1, W_2, ..., W_{lN} be the probabilities that one, two, ... , lN molecules of L are found in a volume V, where N is the Avogadro number. Then,

$$W_1 \propto V,$$

$$W_2 \propto V^2,$$

and, by extension of this reasoning,

$$W_{lN} \propto V^{lN}.$$

W_{lN} is thus the probability that l moles of L will be found in volume V, but the volume containing l moles is inversely proportional to the concentration, c_L. Thus,

$$W_{lN} \propto \left(\frac{1}{c_L} \right)^{lN}.$$

The probability, W_{mN}, that m moles of M should be found in this same volume can similarly be related to the concentration, c_M, as follows:

$$W_{mN} \propto \left(\frac{1}{c_M}\right)^{mN}.$$

The probability that l moles of L and m moles of M will exist in this same volume is then equal to

$$W_{lN}.W_{mN} \propto \left(\frac{1}{c_L}\right)^{lN}.\left(\frac{1}{c_M}\right)^{mN}.$$

Owing to the intermolecular forces, it is more accurate to use activities rather than concentrations, so that, with the usual nomenclature,

$$\text{probability} \propto \left(\frac{1}{a_L}\right)^{lN}.\left(\frac{1}{a_M}\right)^{mN}.$$

This reasoning can be extended to all the products of the reaction to give

$$\text{probability} = W \propto \left(\frac{1}{a_L}\right)^{lN}.\left(\frac{1}{a_M}\right)^{mN} \cdots$$

The corresponding entropy, S_p, is given by

$$S_p = k \ln W$$
$$= \text{constant} - R \ln a_L^l - R \ln a_M^m - \cdots$$
$$= \text{constant} - R \ln (a_L^l . a_M^m \cdots).$$

By similar reasoning, the entropy, S_r, of the reagents is given by

$$S_r = \text{constant} - R \ln (a_E^e . a_F^f \cdots).$$

Let us now suppose that $e.dn$ moles of E and equivalent quantities of the other reagents react. The change in entropy, dS, will be given by

$$dS = (S_p - S_r).dn$$

$$= \left[\text{constant} - R\ln\left(\frac{a_L^l.a_M^m...}{a_E^e.a_F^f...} \right) \right].dn.$$

At constant pressure, the heat of reaction is equal to the enthalpy change. For the conversion of e moles of E and corresponding quantities of the other reagents, let the increase in enthalpy be ΔH. Then, at equilibrium and temperature T,

$$\left[\text{constant} - R\ln\left(\frac{a_L^l.a_M^m...}{a_E^e.a_F^f...} \right) \right].dn = \frac{\Delta H.dn}{T}.$$

Hence,

$$\text{constant} - R\ln\left(\frac{a_L^l.a_M^m...}{a_E^e.a_F^f...} \right) = \frac{\Delta H}{T} \qquad (7.3)$$

If the very slight temperature dependence of ΔH is ignored, then it follows that, for any given temperature, only one value of the expression

$$\frac{a_L^l.a_M^m...}{a_E^e.a_F^f...}$$

will lead to equilibrium. This expression is generally called the equilibrium constant, and is denoted by the symbol K. Equation (7.3) may, therefore, be expressed by the two equations below:

$$K = \frac{a_L^l.a_M^m...}{a_E^e.a_F^f...}; \qquad (7.4)$$

$$\text{constant} - R\ln K = \frac{\Delta H}{T} \qquad (7.5)$$

Equation (7.4) may also be derived from kinetic considerations.

The Van't Hoff Isochore

The value of the equilibrium constant is temperature dependent, and its variation with this parameter is adequately expressed by equation (7.5). Although this is the most useful form of the relationship provided that the temperature range is sufficiently narrow to make ΔH essentially constant, the variation is more commonly represented in books by an expression obtained from the partial differentiation of equation (7.5) at constant pressure. The equation now becomes

$$\left(\frac{\partial}{\partial T}(\ln K)\right)_P = \frac{\Delta H}{RT^2}.$$

This equation is known as the van't Hoff Isochore.

Kirchhoff's Law

Reference has been made at several points to the fact that the quantity of heat absorbed during a chemical or physical change is temperature dependent. Although this slight variation with temperature can be ignored for many purposes, it does, nevertheless, exist, and its nature must now be investigated further.

Suppose that the enthalpy of a set of reagents is H_0 at some temperature and H at a temperature higher by an amount T (in Kelvin), both enthalpies being referred to the same arbitrarily chosen zero. Let the total thermal capacity of the reagents at constant pressure be C_P. Since an increase in enthalpy is equal to the heat absorbed at constant pressure, it is possible to write

$$H = H_0 + \int_0^T C_P.\mathrm{d}T.$$

The second term on the right-hand side cannot be equated with $C_P T$, since C_P will not be constant over a wide range of temperature.

If the reagents now react to form products for which the corresponding parameters are H_0^*, H^* and C_P^*, then

$$H^* = H_0^* + \int_0^T C_P^* . \mathrm{d}T.$$

The heat absorbed during the reaction at constant pressure is equal to the increase in enthalpy, ΔH, so that

$$\Delta H = H^* - H$$

$$= H_0^* - H_0 + \int_0^T \left(C_P^* - C_P \right) \mathrm{d}T.$$

Hence,

$$\left(\frac{\partial}{\partial T} (\Delta H) \right)_P = C_P^* - C_P.$$

The quantity $(C_P^* - C_P)$ representing the difference in thermal capacity between products and reagents is usually written as ΔC_P. Thus,

$$\left(\frac{\partial}{\partial T} (\Delta H) \right)_P = \Delta C_P \qquad (7.6)$$

This statement is known as Kirchhoff's Law.

In order for the law to be used, it is necessary to take account of the variation of ΔC_P with temperature. This is best accomplished if ΔC_P is expressed as a series of the form

$$\Delta C_P = a_0 + a_1 T + a_2 T^2 + a_3 T^3 + ...,$$

where a_0, a_1 etc. are suitable constants, the values of which may be established by thermal capacity measurements at different temperatures. Equation (7.6) then becomes

$$\left(\frac{\partial}{\partial T} (\Delta H) \right)_P = a_0 + a_1 T + a_2 T^2 + a_3 T^3 + ...$$

Hence, by integration

$$\Delta H = I + a_0 T + \frac{1}{2} a_1 T^2 + \frac{1}{3} a_2 T^3 + \frac{1}{4} a_3 T^4 + ...,$$

where I is the integration constant. The measurement of ΔH at one temperature serves to establish the value of I, and makes it possible to calculate ΔH at other temperatures.

Changes of Molecular Weight on Solvent Extraction

The partitioning of a material between two solvents has already been discussed, and it was shown that the ratio of the concentrations is constant at a given temperature, and is, indeed, equal to the ratio of the solubilities. The proviso was, however, made that this law holds good only if the chemical nature of the solute does not differ in the two solvents. The purpose of the present section is to illustrate the considerations that must be made if this proviso is inapplicable.

Consider a material A that remains unchanged in the first solvent, but, in the second, is largely associated to B in accordance with the equation

$$2A \rightleftarrows B.$$

Let the equilibrium constant of this reaction be K, the partition coefficient of A be K_A, the concentrations of A in the two solvents be c_{1A} and c_{2A}, and that of B in the second solvent be c_{2B}.
Then,

$$K = \frac{c_{2B}}{c_{2A}^2},$$

so that

$$c_{2A} = \sqrt{\frac{c_{2B}}{K}}.$$

Also,

$$K_A = \frac{c_{2A}}{c_{1A}}$$

$$= \frac{1}{c_{1A}}\sqrt{\frac{c_{2B}}{K}}.$$

Therefore,

$$KK_A^2 = \frac{c_{2B}}{c_{1A}^2} = \text{constant.}$$

As before, the use of concentrations represents an approximation, and activities would give greater accuracy.

8

Electrochemical Phenomena

Standard Electrode Potentials

Consider a metal M of valency n, and let this be immersed in a solution containing its own ions with an activity equal to a. There may now be a tendency for further ions to leave the solid, but each ion that so detaches itself will leave n electrons behind. In this way, as ions are lost, the solid will acquire a negative charge, which will cause it to bind positive ions more strongly, so that an equilibrium will be established, the metal possessing a certain negative potential, and further loss of positive ions being thereby prevented. Alternatively, the positive ions in the solution may show a tendency to deposit themselves on the solid metal, which they will invest with a positive charge. In this way, an equilibrium will again be attained, the metal possessing a positive potential, which will repel any other positive ions.

In practice, both of these effects will occur simultaneously, and the nature of the metal together with the concentration of the solution will determine their relative importance. In this way, every metal immersed in a solution of its own ions will acquire an electrical potential.

Electrical potential is a quantity that has no absolute value; only differences of potential can be measured. It is necessary, therefore, to select some accurately reproducible system and arbitrarily to assign to its potential a value of zero. Normally, the standard hydrogen electrode fulfils this function. Since it is a rather cumbersome piece of equipment, however, experimental measurements are usually carried out with a more convenient instrument, such as a calomel electrode, the potential of which is accurately known. For descriptions of these and other electrodes, the reader is referred to textbooks of physical chemistry.

For the arguments given below, the nature of the arbitrarily chosen zero of potential is unimportant, as long as it is the same in each case.

Consider once again the metal M of valency n in equilibrium with a solution of its own ions of activity a. In accordance with previous reasoning, the entropy of one mole of ions in solution is given by

$$\text{constant} - R \ln a.$$

The entropy of one mole of atoms in the solid state is constant, so that the change of entropy, ΔS, on dissolution of one mole also takes the form

$$\Delta S = \text{constant} - R \ln a.$$

If the process of dissolution is accompanied by absorption of a quantity q of heat, then, at a temperature T, the dissolution of dm mole at equilibrium may be represented by the equation

$$dS = (\text{constant} - R \ln a).dm = \frac{q.dm}{T} \qquad (8.1)$$

The quantity q must now be examined more closely. Since the process is isothermal, no heat is required to change the temperature, so that all the heat absorbed is used to perform the work that is done. The work required to overcome crystal forces and to counter the osmotic pressure of the solution may be represented as $-w$ per mole, but this is partly offset by the work done as a result of the deposition of electrons on a metal with a potential of E. If the faraday be taken as F, then the dissolution of one mole of ions of valency n liberates a charge of nF. When this is deposited on a surface at a potential E, the work done is equal to nFE. Since it is conventional to regard positive potentials as $+$ and negative ones as $-$, it follows that nFE represents work done by the system, while $-w$ denotes work done on the system. Accordingly,

$$q = -w - nFE.$$

When this is substituted into equation (8.1), the result is

$$\left(\text{constant} - R\ln a\right).dm = \frac{\left(-w - nFE\right).dm}{T}.$$

Hence,

$$\text{constant} - R\ln a = \frac{-w - nFE}{T} \qquad (8.2)$$

Consider the same metal now to be immersed in a solution of different concentration, so that the activity of the dissolved ions is a_1 and the potential of the metal is E_1. Then,

$$\text{constant} - R\ln a_1 = \frac{-w_1 - nFE_1}{T}, \qquad (8.3)$$

where w_1 represents the new value for the quantity previously called w.

By subtraction of (8.2) from (8.3), the following result is obtained:

$$R\ln\frac{a}{a_1} = \frac{\left(w - w_1\right) + nF\left(E - E_1\right)}{T} \qquad (8.4)$$

The quantity $(w - w_1)$ is very small, while F is very large, so that equation (8.4) may be approximated as follows:

$$R\ln\frac{a}{a_1} = \frac{nF\left(E - E_1\right)}{T}. \qquad (8.5)$$

Let the potential of a metal immersed in a solution of unit activity be E_0. Then, if

$$a_1 = 1$$

and

$$E_1 = E_0,$$

Equation (8.5) becomes

93

$$R \ln a = \frac{nF(E - E_0)}{T},$$

which is more usually written as

$$E = E_0 + \frac{RT}{nF} \ln a.$$

This statement is known as the Nernst equation. The quantity E_0, which is virtually independent of temperature, is referred to as the standard electrode potential of the metal concerned.

In the use of the Nernst equation, careful attention must be given to the question of units. Obviously, the two potentials E and E_0 must be measured in the same units and referred to the same arbitrary zero. Normally, they are quoted in volts relatively to the standard hydrogen electrode.

The quantity RT has the dimension of energy per mole, and F has that of electric charge per mole. If, therefore, RT/F is to be in volts, then R must be measured in $Jmol^{-1} K^{-1}$, T in K and F in coulombs mol^{-1}. The units of a have already been discussed in Chapter 6, and n is a dimensionless quantity.

At one time, an unfortunate dichotomy of sign conventions existed between the two sides of the Atlantic. The convention used above is that which developed in Europe, while American scientists quoted standard electrode potentials with opposite signs. If this procedure is adopted, then the Nernst equation must be written as

$$E = E_0 - \frac{RT}{nF} \ln a.$$

Nowadays, most workers have adopted the European convention, but confusion still occurs when older texts or journals are consulted. According to present practice, the standard electrode potentials of the alkali metals are negative, and those of the noble metals positive. As a result, the Nernst Equation is written as

$$E = E_0 + \frac{RT}{nF} \ln a.$$

The use of tables of natural logarithms is markedly less convenient than that of logarithms to the base 10. Although electronic calculators and computers have largely obviated this difficulty, it remains instructive to examine an alternative form of the Nernst equation. It is easily proved that

$$\ln a = \log_{10} a . \log_e 10.$$

Hence, in the last term of the Nernst equation, logarithms to the base 10 may be used, provided that the term is also multiplied by

$$\log_e 10 = 2.303.$$

Since, however, R and F are constants, they can be combined with this quantity to produce a single factor. Furthermore, most experimental measurements of electrode potentials are carried out in thermostats at 25°C, i.e. at 298.15 K. This figure can also be incorporated in the new factor, so that the Nernst equation emerges as

$$E = E_0 + \frac{0.059}{n} \log_{10} a$$

at 25°C.

From this, it can be seen that a tenfold increase in the activity of the ions in solution results in a rise of $0.059/n$ volt in the potential of the electrode. A 100-fold increase in activity raises the potential by $2 \times 0.059/n$ volt, and so on.

Standard electrode potentials and the Nernst equation determine which metal will displace another from solution. They also have many other practical applications such as in concentration cells, ion selective electrodes, measurement of solubility products etc. For these, however, the reader must refer to the appropriate textbooks.

Redox Reactions

Reactions in which one reagent is oxidised while another is reduced are called redox reactions. Since all oxidations and reductions can be regarded as respectively losses and gains of electrons, it follows that they can be treated in a manner similar to that employed for the ionisation of metals. Indeed, the formation of ions from neutral atoms of a metal is merely a special case of oxidation.

Consider a reagent that is changed from its oxidised form, written as Ox, to its reduced form, denoted by Red, by the gain of n electrons per molecule or ion. Let the activity of the oxidised form be a_{Ox} and that of the reduced form a_{Red}. It follows that the increase in entropy, dS, on conversion of dm mole is given by

$$dS = (\text{constant} - R \ln a_{Red} + R \ln a_{Ox}).dm$$

$$= \left(\text{constant} + R \ln \frac{a_{Ox}}{a_{Red}} \right).dm.$$

The necessary electrons must be derived from some source. Let this be an electrode of chemically inert material with a potential of E. The work that must be done to remove $nF.dm$ coulombs from this might at first sight seem to be equal to $nFE.dm$, but actually there will be an additional term of $nFw.dm$, the precise value of which depends on the nature of the electrode. Accordingly, at equilibrium,

$$dS = \left(\text{constant} + R \ln \frac{a_{Ox}}{a_{Red}} \right).dm$$

$$= \frac{nFw.dm + nFE.dm}{T}.$$

From this, it follows that

$$\text{constant} + R \ln \frac{a_{Ox}}{a_{Red}} = \frac{nFw + nFE}{T} \tag{8.6}$$

If the activities of Ox and Red are equal, let the value of E be E_0. Equation (8.6) then becomes

$$\text{constant} = \frac{nFw + nFE_0}{T} \tag{8.7}$$

The two constant terms on the left of equations (8.6) and (8.7) are equal, so that subtraction of (8.7) from (8.6) yields the result

$$R \ln \frac{a_{Ox}}{a_{Red}} = \frac{nF(E - E_0)}{T},$$

which may be written as

$$E = E_0 + \frac{RT}{nF} \ln \frac{a_{Ox}}{a_{Red}}.$$

This is also known as the Nernst equation, and the quantity E_0 is variously referred to as the redox potential or the oxidation potential.

All the comments previously made about units are still apposite, and, if measurements are carried out at 25°C, this new Nernst equation can also be written as

$$E = E_0 + \frac{0.059}{n} \log_{10} \frac{a_{Ox}}{a_{Red}}.$$

The purpose of redox potentials can best be illustrated by an example. Consider a solution that contains the ions Fe^{+++}, Fe^{++}, Sn^{++++} and Sn^{++}. Although most compounds of quadrivalent tin are either covalent or complexed forms of this metal, the ion Sn^{++++} will be used here for simplicity. The solution now contains two redox systems represented by the following two equations:

$$Sn^{++++} + 2e \rightleftarrows Sn^{++}$$

$$Fe^{+++} + e \rightleftarrows Fe^{++}$$

If the redox potentials of these two systems are E_{Sn} and E_{Fe} respectively, then the Nernst equations can be written as follows, the obvious nomenclature being used for activities:-

$$E = E_{Sn} + \frac{RT}{2F} \ln \frac{a_{Sn^{++++}}}{a_{Sn^{++}}} \qquad (8.8)$$

97

$$E = E_{Fe} + \frac{RT}{F} \ln \frac{a_{Fe^{+++}}}{a_{Fe^{++}}} \tag{8.9}$$

Equation (8.9) can be rewritten as

$$E = E_{Fe} + \frac{RT}{2F} \ln \frac{a_{Fe^{+++}}^2}{a_{Fe^{++}}^2} \tag{8.10}$$

If the whole system is at equilibrium, then it will have undergone the reversible reaction

$$Sn^{++} + 2Fe^{+++} \rightleftarrows Sn^{++++} + 2Fe^{++}.$$

The equilibrium must, however, be electrical as well as chemical, so that the values of E in equations (8.8) and (8.10) must be equal. It therefore follows that

$$E = E_{Sn} + \frac{RT}{2F} \ln \frac{a_{Sn^{++++}}}{a_{Sn^{++}}} = E_{Fe} + \frac{RT}{2F} \ln \frac{a_{Fe^{+++}}^2}{a_{Fe^{++}}^2}.$$

Hence,

$$E_{Fe} - E_{Sn} = \frac{RT}{2F} \ln \left(\frac{a_{Sn^{++++}} . a_{Fe^{++}}^2}{a_{Sn^{++}} . a_{Fe^{+++}}^2} \right) \tag{8.11}$$

The quantity

$$\frac{a_{Sn^{++++}} . a_{Fe^{++}}^2}{a_{Sn^{++}} . a_{Fe^{+++}}^2}$$

is equal to the equilibrium constant, K, of the reversible reaction, so that equation (8.11) can be written as

$$E_{Fe} - E_{Sn} = \frac{RT}{2F} \ln K.$$

Thus, it can be seen that the difference between the two redox potentials determines the value of the equilibrium constant. Because of the logarithmic nature of the relationship, even a small change in the difference between potentials occasions an enormous change in the value of K. Thus, from a mere inspection of a list of redox potentials, it is possible to deduce whether one system oxidises or reduces another. It must, however, be pointed out emphatically that redox potentials serve only to indicate the final state of equilibrium. They give no indication of the rate at which this equilibrium will be attained. Thus, it is possible for a mixture to be thermodynamically unstable, i.e. not at equilibrium, but kinetically stable, i.e. not changing at a perceptible rate.

Once again, two different sign conventions developed at first. That which is explained above is the European convention, in which the Nernst equation is written as

$$E = E_0 + \frac{RT}{nF} \ln \frac{a_{Ox}}{a_{Red}},$$

and all powerful oxidising agents have positive redox potentials, while those of powerful reducing agents are negative. This is the form that has now been adopted almost universally, but an older American convention connected the two terms on the right-hand side of the equation with a minus and reversed the signs of the redox potentials.

Finally, it must be mentioned that some redox systems involve not only the oxidised and reduced forms of the reagent, but water and hydrogen or hydroxide ions as well. Since the water is present in great excess, its activity may be regarded as constant, but the hydrogen or hydroxide ions must be included in the Nernst equation, which thus becomes a little more complicated. Consider, for example, the reduction of permanganate ions to bivalent manganese in acid solution:

$$MnO_4^- + 8H^+ + 5e \rightleftarrows Mn^{++} + 4H_2O.$$

By the reasoning given above, the Nernst equation will be seen to be

$$E = E_0 + \frac{RT}{5F} \ln\left(\frac{a_{MnO_4^-} \cdot a_{H^+}^8}{a_{Mn^{++}}} \right).$$

As can easily be verified, the use of the redox potential of this system for the calculation of equilibrium constants is unaffected.

9

Heat Engines

The Carnot Cycle

The earliest steam engines and internal combustion engines all worked on the reciprocating basis, that is to say a gas known as the working substance was compressed in a cylinder and allowed to expand. It was this expansion that accomplished the work required of the engine, but, in order to return the piston to its original position, the working substance had to be compressed again. For this purpose, work had to be done on the working substance, and it is obvious that the work done on the engine during the compression had to be less than that which was done by it during the expansion. Although the mechanics of a turbine are more difficult to explain concisely, the same considerations hold in principle.

During the expansion, when the engine is doing work, energy must be taken in by some process such as the combustion of fuel. During the compression, when work is done on the engine, energy in the form of heat is given out. If the heat energy taken in is denoted by Q_1 and that given out by Q_2, then it might be thought that the difference between these two quantities would be the mechanical work done. In practice, however, this is not the case, since a certain amount of energy will be lost. The principal sources of wastage include friction in the moving parts and heat losses due to contact between hot surfaces and colder ones. Careful engineering will serve to minimise, though not to eliminate, these losses, but, for theoretical purposes, it is possible to postulate an engine in which wastage has been avoided completely, so that maximum efficiency is achieved. In such a case,

$$\text{work done by engine} = W = Q_1 - Q_2.$$

The efficiency of an engine is defined as the output of mechanical work divided by the original input of energy, so that

$$\text{efficiency} = \frac{W}{Q_1} = \frac{Q_1 - Q_2}{Q_1}.$$

For simplicity, let us consider the expansion to occur isothermally at a temperature of T_1, the volume changing from V_1 to V_2. In this case, if the working substance approximates to an ideal gas,

$$\text{work done by machine} = RT_1 \ln \frac{V_2}{V_1}$$

per mole of working substance.

Clearly, the compression will have to start with the final volume achieved on expansion, and return to the original one. This will mean, however, that all the work performed on expansion will have to be returned to the engine during the course of an isothermal compression, unless the latter occurs at a lower temperature T_2. Since a change of temperature must not be achieved by mere leakage of heat, it follows that an adiabatic expansion will have to be interposed to lower the temperature from T_1 to T_2, and then, after the isothermal compression, an adiabatic compression will return the temperature to T_1. The four stages of the cycle may now be summarised as follows:-

I. Isothermal expansion at temperature T_1.
Heat absorbed $= Q_1$.
II. Adiabatic expansion from temperature T_1 to T_2.
III. Isothermal compression at temperature T_2.
Heat evolved $= Q_2$.
IV. Adiabatic compression from temperature T_2 to T_1. Original volume reached.

The precise construction of the engine need not concern us, but one other point must be made. The engine absorbs a quantity of heat, Q_1, during stage I, turns some of it into mechanical work, W, and rejects the remainder, Q_2, during stage III. Since it is postulated that there are no friction losses or leakages of heat, it should be possible to work the engine in reverse, that is to say to absorb a quantity of heat, Q_2, at the lower temperature, T_2, augment it with a quantity of work, W, performed on the engine from some outside source, and reject a larger amount of heat, Q_1, at a higher temperature, T_1. In other words, the device should be capable of acting as a heat engine or as a refrigerator. If, however, friction losses or adventitious heat losses occur, this reversibility will be lost.

At the end of each cycle, the contents of the cylinder will be in exactly the same state as they were before the cycle began. The same, however, cannot be said of the environment. This will have lost a quantity of heat, Q_1, during the isothermal expansion and recovered a smaller quantity, Q_2, during the isothermal compression. If irreversible losses of heat are to be avoided, then that part of the environment that supplies or recovers the heat must be at the same temperature as the engine. Accordingly,

$$\text{Increase in entropy of environment} = \frac{Q_2}{T_2} - \frac{Q_1}{T_1}.$$

Since the contents of the cylinder undergo no overall change in state, and hence in entropy, the change in the environment represents that in the whole of an isolated system, so that

$$\frac{Q_2}{T_2} - \frac{Q_1}{T_1} \geq 0,$$

where the equality holds only if the engine is reversible, and the inequality if it is irreversible.

It is now possible to rearrange the expression above as follows:-

$$\frac{Q_2}{T_2} \geq \frac{Q_1}{T_1}.$$

$$\frac{Q_2}{Q_1} \geq \frac{T_2}{T_1}.$$

$$-\frac{Q_2}{Q_1} \leq -\frac{T_2}{T_1}.$$

$$1 - \frac{Q_2}{Q_1} \leq 1 - \frac{T_2}{T_1}.$$

$$\frac{Q_1 - Q_2}{Q_1} \leq \frac{T_1 - T_2}{T_1}.$$

The left-hand side of this last equation gives the efficiency of the engine, so that we may write

$$\text{efficiency} \leq \frac{T_1 - T_2}{T_1}.$$

Hence, the most efficient type of heat engine that may be postulated is one that is perfectly reversible, but even the efficiency of this is limited to $(T_1 - T_2)/T_1$. Any irreversible heat engine is necessarily less efficient.

The theoretically postulated reversible heat engine described above is known as a Carnot Cycle. It may be envisaged as a cylinder closed with a frictionless piston, the walls of the cylinder being of perfectly insulating material except at one point, to which may be attached a heat source at temperature T_1, a heat sink at temperature T_2 or a perfect insulator that enables adiabatic processes to occur within the cylinder.

Finally, the question of reversibility requires a little elucidation. Even a practical heat engine with all its mechanical imperfections can be designed to function in reverse as a refrigerator. This does not, however, make it reversible in the thermodynamic sense, since the amounts of heat involved would be different. The energy losses when the engine is driven in one direction would not appear as energy gains when the engine is reversed. Instead, they would be augmented by further friction and heat losses.

An Alternative Approach

In the previous section, the efficiency of a heat engine was derived from considerations of entropy. Historically, however, this is not the way in which such knowledge was acquired. At the time when

Carnot first derived his theorem, entropy was as yet unknown, and it is interesting to trace the essential stages of what is usually called classical thermodynamics.

Let us again consider the four stages of the Carnot Cycle for one mole of working substance:-

I. Isothermal expansion from volume V_1 to V_2 at temperature T_1.

II. Adiabatic expansion from volume V_2 to V_3, causing a temperature change from T_1 to T_2.

III. Isothermal compression from volume V_3 to V_4 at temperature T_2.

IV. Adiabatic compression from volume V_4 to V_1, causing a temperature change from T_2 to T_1.

The rules of adiabatic changes may now be applied to stages II and IV with the following results:-

$$\frac{T_1}{T_2} = \left(\frac{V_3}{V_2}\right)^{(\gamma-1)} = \left(\frac{V_4}{V_1}\right)^{(\gamma-1)}.$$

Hence,

$$\frac{V_3}{V_2} = \frac{V_4}{V_1}.$$

Therefore,

$$\frac{V_4}{V_3} = \frac{V_1}{V_2}.$$

The four stages may now be rewritten as described below. In each case, one mole of working substance is being considered.

I. Work done by engine = heat absorbed by engine

$$= Q_1 = RT_1 \ln\frac{V_2}{V_1}.$$

II. Work done by engine

$$= C_V(T_1 - T_2),$$

where C_V is the molar thermal capacity at constant volume.

III. Work done by engine = heat absorbed by engine

$$= -Q_2 = RT_2 \ln \frac{V_4}{V_3}$$

$$= RT_2 \ln \frac{V_1}{V_2}$$

$$= -RT_2 \ln \frac{V_2}{V_1}.$$

IV. Work done by engine

$$= C_V(T_2 - T_1)$$
$$= - C_V(T_1 - T_2).$$

From these relationships, the following conclusions may be drawn:-

Total work done by engine $= W$

$$= Q_1 + C_V(T_1 - T_2) - Q_2 - C_V(T_1 - T_2)$$
$$= Q_1 - Q_2$$

$$= RT_1 \ln \frac{V_2}{V_1} - RT_2 \ln \frac{V_2}{V_1}$$

$$= (T_1 - T_2) R \ln \frac{V_2}{V_1}.$$

$$\text{Efficiency} = \frac{W}{Q_1}$$

$$= \frac{Q_1 - Q_2}{Q_1}$$

$$= \frac{(T_1 - T_2)R \ln \dfrac{V_2}{V_1}}{RT_1 \ln \dfrac{V_2}{V_1}}$$

$$= \frac{T_1 - T_2}{T_1}.$$

This may be summarised as

$$\text{efficiency} = \frac{Q_1 - Q_2}{Q_1} = \frac{T_1 - T_2}{T_1}.$$

Up to this point, the calculation is concerned only with the Carnot Cycle. It can, however, readily be proved that no heat engine can have an efficiency exceeding that of the Carnot Cycle. Let us suppose that a more efficient heat engine existed, and that it drew a quantity of heat, Q_1^*, from a source at temperature T_1, converted some of it into mechanical work, W, and returned the remainder, Q_2^*, to a sink at temperature T_2. Since the Carnot Cycle is reversible, the more efficient engine could be used to drive a Carnot Cycle in reverse. The latter would then draw a quantity of heat, Q_2, from the reservoir at temperature T_2, add to it an amount of work, W, and reject an amount of heat, Q_1, at temperature T_1. The following relationships then hold good:-

$$W = Q_1^* - Q_2^* = Q_1 - Q_2.$$

$$\frac{W}{Q_1^*} > \frac{W}{Q_1}.$$

Therefore,

$$\frac{1}{Q_1^*} > \frac{1}{Q_1}.$$

$$Q_1^* < Q_1$$

and

$$Q_2^* < Q_2.$$

Thus, it can be seen that the amount of heat transferred from the higher temperature to the lower one would be less than that transferred in the opposite direction. Although entropy was unknown at the time when the Carnot Cycle was first propounded, the Second Law of Thermodynamics was familiar as an experimental fact. The use of a more efficient engine to drive a Carnot Cycle in reverse would lead to an overall transfer of heat from a lower temperature to a higher one. This is in breach of the Second Law, so that it may be taken that no heat engine more efficient than a Carnot Cycle can exist.

This proposition is capable of a different interpretation. If the Second Law could be breached by means of a heat engine more efficient than the Carnot Cycle, then the energy transferred to a higher temperature could be used to drive another engine, and a perpetual motion machine would thus have been achieved. Since no energy would have to be created for this, the result is described as perpetual motion of the second kind. If it existed, it would amount to a recycling of energy already used.

The same considerations would apply if a Carnot Cycle could be used to drive in reverse a less efficient, but reversible, heat engine. Thus, it may be seen that no reversible heat engine can be less efficient than a Carnot Cycle. Since it cannot be more efficient either, it must clearly have the same efficiency. The friction losses, heat losses and other mechanical imperfections that prevent a heat engine from being reversible obviously amount to an impairment of efficiency, so that the following may be said:-

For any heat engine,

$$\frac{Q_1 - Q_2}{Q_1} \leq \frac{T_1 - T_2}{T_1},$$

where the equality applies to a reversible heat engine, and the inequality to an irreversible one.

This expression leads to the following conclusions:-

$$1 - \frac{Q_2}{Q_1} \leq 1 - \frac{T_2}{T_1}.$$

$$-\frac{Q_2}{Q_1} \leq -\frac{T_2}{T_1}.$$

$$\frac{Q_2}{Q_1} \geq \frac{T_2}{T_1}.$$

$$\frac{Q_2}{T_2} \geq \frac{Q_1}{T_1}.$$

$$\frac{Q_2}{T_2} - \frac{Q_1}{T_1} \geq 0.$$

It was from this last expression that the idea of entropy was originally derived. Entropy, S, was at first seen purely as an experimentally measurable quantity expressed in terms of heat absorbed and temperature. Its connection with randomness, W, and the equation

$$S = k \ln W$$

were derived much later by Boltzmann, in whose honour the constant k was named.

These considerations offer not only a different approach to thermodynamics but also a further insight into the working of engines. It will be seen that every heat engine transfers heat from a high temperature to a lower one. Furthermore, it is difficult to devise any machine that does not ultimately depend on a heat engine. Although an electric motor does not work on these principles, the generation of the electricity does. A steam or diesel generator is obviously a heat engine, but, less obviously, even hydroelectric or wind-driven generators depend ultimately on the transfer of heat from the high temperature of the sun to much lower terrestrial ones. The same may be said of machines driven manually by human beings or pulled by animals. An electric battery depends on the conversion of a very reactive metal into one of its compounds. The energy released is a manifestation of the greater energy contained in the free metal. Such a reactive substance, however, normally occurs naturally in the form of one of its compounds. For the free metal to be formed, energy must be imparted to the ore either by a smelting process at high temperature or by an electrolytic process using electricity generated by a heat engine. Thus, it may be seen that almost all machines are, in fact, heat engines, which transfer energy from a high temperature to a lower one. The fact that this energy cannot rise spontaneously to a higher temperature makes it impossible to construct a machine giving perpetual motion of the second kind.

Heat Pumps

As has been explained above, a refrigerator is actually a heat engine working in reverse; that is to say, it absorbs heat from a source at low temperature, adds to it a quantity of mechanical work supplied by some other engine, and rejects the sum of the two at a higher temperature. Since the engine providing the mechanical work absorbs heat at a high temperature and rejects the excess at a lower temperature, the net effect is the transfer of heat from a high temperature to a lower one, so that the requirements of the Second Law are satisfied.

There remains, however, the possibility of using a refrigerator in such a way that the purpose is not to cool the inside, but to heat the outside. In particular, it has been suggested that such a device, known as a heat pump, could be used to withdraw heat from the atmosphere and use it to warm a building. Calculations suggest that this would furnish a more efficient method of heating than conventional systems. The practical difficulties in implementing the idea seem to be considerable, but it nevertheless remains at least theoretically feasible.

10

Free Energy and Chemical Potential

Gibbs Free Energy and Helmholtz Free Energy

In earlier chapters, chemical and physical equilibria as well as elec-trochemical phenomena were discussed as consequences of the properties of entropy. That is, indeed, the fundamental way of regarding these matters, but an alternative method, commonly employed and frequently useful, makes use of Free Energy.

Consider a part of a system not doing electrical work. Let the part have entropy S, internal energy U, enthalpy H and temperature T measured in Kelvin.

If the part under consideration is at constant pressure, then

$$\text{heat absorbed } = \text{d}H.$$

Therefore,

$$\text{d}S \geq \frac{\text{d}H}{T},$$

where the equality denotes a reversible absorption of heat, and the inequality an irreversible one. Then it follows that

$$T.\text{d}S \geq \text{d}H.$$

Therefore,

$$\text{d}H - T.\text{d}S \leq 0.$$

If the amount of heat absorbed is so small as to leave the tem-perature effectively unchanged, then

$$\text{d}(H - TS) \leq 0.$$

113

The quantity $(H - TS)$ is known as the Gibbs Free Energy, and is generally denoted by G. Hence,

$$G = H - TS$$

and

$$dG \leq 0,$$

where the equality again refers to a reversible process, and the inequality to an irreversible one.

Consider the same process to be carried out at constant volume. Then,

$$\text{heat absorbed} = dU.$$

Therefore,

$$dS \geq \frac{dU}{T}.$$

$$T.dS \geq dU.$$

$$dU - T.dS \leq 0$$

and

$$d(U - TS) \leq 0.$$

The quantity $(U - TS)$ is known as the Helmholtz Free Energy, and is accorded the symbol A. Thus,

$$A = U - TS$$

and

$$dA \leq 0,$$

where the equality and inequality signs have the same connotations as before.

Properties of Gibbs Free Energy

A change in Gibbs Free Energy can be written as follows:-

$$G = H - TS$$
$$= U + PV - TS,$$

where P is the pressure and V is the volume. If electrical work is done,

$$dG = dq - dw_P - dw_e + P.dV + V.dP - T.dS - S.dT,$$

where dq is the heat absorbed by the part of the system under consideration, dw_P is the work done by it against the pressure and dw_e is the electrical work done by it. Since

$$P.dV = dw_P,$$

it follows that

$$dG = dq - dw_e + V.dP - T.dS - S.dT.$$

Two cases may now be distinguished. Both assume that the part of the system under consideration is closed to the ingress or egress of matter, though not energy.

Firstly, if the part of the system is at constant temperature and pressure, then

$$dG = dq - dw_e - T.dS.$$

If, moreover, the change is reversible, then

$$T.dS = dq,$$

so that

$$dG = - dw_e.$$

Thus, it may be said that, in a part of a system closed to the ingress or egress of matter and undergoing a reversible change at constant temperature and pressure, the increase in Gibbs Free Energy is equal to minus the electrical work done by the system.

Alternatively, let us consider a part of the system that does no electrical work. Then

$$dG = dq + V.dP - T.dS - S.dT.$$

If the change is reversible,

$$dG = V.dP - S.dT.$$

From this, two other useful equations may be derived, namely that

$$\left(\frac{\partial G}{\partial P}\right)_T = V$$

and

$$\left(\frac{\partial G}{\partial T}\right)_P = -S.$$

The former of these will be required for the discussion of chemical potential, and the latter for the Gibbs–Helmholtz equation.

Properties of Helmholtz Free Energy

Since chemical reactions are more commonly made to occur at constant pressure than at constant volume, the Helmholtz Free Energy is used less frequently than its Gibbs counterpart. Indeed, when reference is made to free energy without further qualification, the allusion is normally to the latter. Nevertheless, this chapter would be incomplete without some description of Helmholtz Free Energy. Changes in this quantity for part of a system not accessible to additional matter may be written as follows:-

$$A = U - TS.$$
$$dA = dq - dw - T.dS - S.dT,$$

116

where dw is the total work done by the part of the system concerned.
At constant temperature,

$$dA = dq - dw - T.dS.$$

If the change is reversible,

$$T.dS = dq,$$

so that

$$dA = - dw.$$

From this, it can be seen that, for a part of a system closed to the ingress or egress of matter and undergoing a reversible change at constant temperature, the increase in Helmholtz Free Energy is equal to minus the total work done by that part of the system.

Actually, the two equations

$$dG = - dw_e$$

and

$$dA = - dw$$

are not as different as they may at first sight appear, since Helmholtz Free Energy is normally used at constant volume, when there is no pressure work, and the electrical work represents the total.

Free Energy as a Condition of State

It has already been explained that the internal energy, enthalpy and entropy of a system all depend on the current state of the system and not on the process by which that state was attained. From the equations

$$G = H - TS$$

and

$$A = U - TS,$$

it is clear that the same must apply to the two forms of free energy.

This fact is of importance in the understanding of reversible and irreversible processes. Consider, for example, a part of a system closed to matter and performing electrical work at constant temperature and pressure. It has already been shown that, in such a case,

$$dG = dq - dw_e - T.dS.$$

Hence,

$$dw_e = dq - T.dS - dG.$$

The only quantities in this equation which depend on the route taken by the change are dw_e and dq. If the change is reversible, then

$$dq = T.dS$$

and

$$dw_e = - dG.$$

If, however, the change is irreversible, then

$$dq < T.dS$$

and

$$dw_e < - dG.$$

In each case, dS and dG remain the same, since they depend only on the initial and final conditions, and not on the process by which the one is converted to the other.

Thus, it may be seen that a reversible process results in the maximum absorption of heat and yields the maximum of electrical work, while the same reaction carried out irreversibly results in lower values for these two quantities. Indeed, a reaction that is endothermic when carried out reversibly may sometimes become exothermic under irreversible conditions. A good example of this is the solution of zinc in dilute sulphuric acid, which is endothermic and yields electrical work when carried out reversibly in a Daniell cell, but is exothermic and yields no electrical work when it is carried out irreversibly by the simple treatment of the metal with acid.

Melting Point and Boiling Point

All the equilibria explained in Chapters 5, 7 and 8 in terms of entropy can also be described in terms of free energy. Space does not permit a discussion of them all, but a few examples will suffice.

Consider the melting or solidification of a substance at constant temperature and pressure. With the usual nomenclature,

$$\Delta G = \Delta H - T . \Delta S.$$

If the substance is melting, then ΔH and ΔS will be positive, and there will be only one temperature for which

$$\Delta G = 0.$$

Melting will then be a reversible process, so that that temperature will be the melting point. At any higher temperature,

$$\Delta G < 0,$$

so that melting will be irreversible. If the substance is solidifying, then ΔH and ΔS will both be negative. ΔG will still be zero at the same temperature, but a negative value of ΔG indicating an irreversible change will now occur at temperatures below the melting point.

Similar considerations apply to the boiling point.

Chemical Potential

In considerations of chemical and physical equilibrium, it is often convenient to think in terms of one mole. The Gibbs Free Energy of one mole of a substance is called the chemical potential, and is generally denoted by the symbol μ.

It was shown above that

$$\left(\frac{\partial G}{\partial P}\right)_T = V.$$

If this equation is applied to one mole of an ideal gas, then

$$\left(\frac{\partial \mu}{\partial P}\right)_T = \frac{RT}{P},$$

and

$$\mu = \mu_0 + RT \ln P, \qquad (10.1)$$

where μ_0 is a constant. The condition that

$$\mu = \mu_0$$

requires that

$$P = 1.$$

In this case, the gas is said to be in its standard state.

The analogy between ideal gases and ideal solutions enables equation (10.1) to be applied to the latter, provided that P now represents the osmotic pressure. This quantity, however, is proportional to the concentration, c, and substitution of the one for the other would merely modify the constant μ_0, so that we may write

$$\mu = \mu_0 + RT \ln c, \qquad (10.2)$$

where μ is the chemical potential of the solute. The standard state of the latter is represented by unit concentration.

In the case of a real gas or a real solution, equations (10.1) and (10.2) become

$$\mu = \mu_0 + RT \ln P^* \qquad (10.3)$$

and

$$\mu = \mu_0 + RT \ln a \qquad (10.4)$$

respectively, where P^* is the fugacity and a is the activity. Unit values of these quantities again give the standard state.

In all these equations, careful attention must be given to units, which must, of course be consistent.

The derivation of Raoult's Law may be regarded as an instructive example of the use of chemical potentials.

Any system with two components is referred to as a binary system, and, when the constituent materials are respectively solvent and solute, the various parameters are distinguished by a subscript 1 for the former and a subscript 2 for the latter. Let the following symbols then be used:

Mol fractions	N_1 and N_2
Vapour pressure of pure solvent	p_1^*
Vapour pressure of solvent in solution	p_1
Chemical potentials	μ_1 and μ_2
Standard chemical potentials	μ_{1s} and μ_{2s}
Temperature	T
Gas constant	R
Total Gibbs Free Energy	G.

If the solution is in equilibrium with the solvent vapour, then the transfer of a small quantity of solvent to the vapour phase should not result in a change of Gibbs Free Energy. Thus,

chemical potential of solvent in liquid phase
= chemical potential of solvent in vapour phase
$$= \mu_1$$
$$= \mu_{1s} + RT \ln p_1.$$

Therefore,

$$d\mu_1 = \frac{RT}{p_1}.dp_1. \qquad (10.5)$$

Now the Gibbs Free Energy of each component of the system is given by the product of the chemical potential and the number of moles present, the total Gibbs Free Energy of the system being the sum of those of the constituents. Let the number of moles of solvent and solute be n_1 and n_2, respectively. Then

$$G = n_1\mu_1 + n_2\mu_2.$$

$$\frac{G}{n_1 + n_2} = N_1\mu_1 + N_2\mu_2.$$

Since

$$dG = 0,$$

$$N_1.d\mu_1 + \mu_1.dN_1 + N_2.d\mu_2 + \mu_2.dN_2 = 0.$$

If the system is closed, then N_1 and N_2 must be constant, so that

$$dN_1 = 0$$

and

$$dN_2 = 0.$$

Hence,

$$N_1.d\mu_1 + N_2.d\mu_2 = 0. \qquad (10.6)$$

Equation (10.6) is of great importance in calculations concerning binary systems, and is called the Gibbs–Duhem equation. Substitution into equation (10.5) yields

$$\frac{RT}{p_1}.dp_1 = -\frac{N_2}{N_1}.d\mu_2,$$

but

$$\mu_2 = \mu_{2s} + RT \ln N_2$$

since

$$N_2 \propto \text{concentration of solute.}$$

Hence,

$$d\mu_2 = \frac{RT}{N_2}.dN_2.$$

Therefore,

$$\frac{RT}{p_1}.dp_1 = -\frac{RT}{N_1}.dN_2.$$

$$\frac{dp_1}{p_1} = -\frac{dN_2}{N_1} \qquad (10.7)$$

In a dilute solution, however, N_1 is not very different from unity, so that equation (10.7) can be written as

$$-dN_2 = \frac{dp_1}{p_1}.$$

By integration,

$$\int_0^{N_2} -dN_2 = \int_{p_1^*}^{p_1} \frac{dp_1}{p_1}.$$

$$-N_2 = \ln\frac{p_1}{p_1^*}$$

$$= \ln\left[1 - \left(1 - \frac{p_1}{p_1^*}\right)\right]$$

$$= \ln\left[1 - \left(\frac{p_1^* - p_1}{p_1^*}\right)\right].$$

It has already been explained that, for any small quantity x,

$$\ln(1+x) \approx x.$$

It follows, therefore, that

$$-N_2 \approx -\left(\frac{p_1^* - p_1}{p_1^*}\right)$$

and

$$N_2 = \frac{p_1^* - p_1}{p_1^*}.$$

This is Raoult's Law.

Reversible Reactions

The existence of equilibrium constants can also be derived by means of Gibbs Free Energy. Consider the following reaction:

$$eE + fF + \ldots \rightleftarrows lL + mM + \ldots$$

If we consider a change of e moles of E and the appropriate number of moles of the other reagents, then

gain of Gibbs Free Energy of products
$$= l\mu_L + RT \ln a_L^l + m\mu_M + RT \ln a_M^m + \ldots$$
$$= l\mu_L + m\mu_M + \ldots + RT \ln (a_L^l.a_M^m\ldots),$$

where the symbols have their obvious meanings. Now

loss of Gibbs Free Energy of reagents
$$= e\mu_E + RT \ln a_E^e + f\mu_F + RT \ln a_F^f + \ldots$$
$$= e\mu_E + f\mu_F + \ldots + RT \ln (a_E^e.a_F^f \ldots).$$

Hence, the total change in Gibbs Free Energy, ΔG, is given by

$$\Delta G = l\mu_L + m\mu_M + ... - \left(e\mu_E + f\mu_F + ...\right) + RT \ln\left(\frac{a_L^l.a_M^m...}{a_E^e.a_F^f...}\right)$$

$$= \Delta G_0 + RT \ln\left(\frac{a_L^l.a_M^m...}{a_E^e.a_F^f...}\right),$$

where

$$\Delta G_0 = l\mu_L + m\mu_M + ... - \left(e\mu_E + f\mu_F + ...\right).$$

It can be seen that ΔG_0 would be the change in Gibbs Free Energy if all the reagents and products were in their standard states. This is clearly a constant. At equilibrium, however,

$$\Delta G = 0.$$

Therefore,

$$\Delta G_0 = -RT \ln\left(\frac{a_L^l.a_M^m....}{a_E^e.a_F^f....}\right).$$

Let

$$\left(\frac{a_L^l.a_M^m....}{a_E^e.a_F^f....}\right) = K.$$

Then

$$\Delta G_0 = -RT \ln K. \qquad (10.8)$$

Thus, it can be seen that K will be constant at a given temperature. It is the equilibrium constant of the reaction, and equation (10.8), which is known as the van't Hoff Isotherm, gives an insight into the reasons for the actual value of this parameter in a given instance.

The Gibbs–Helmholtz Equation and the Van't Hoff Isochore

Gibbs Free Energy was defined above by the equation

$$G = H - TS.$$

It was, however, also proved that

$$\left(\frac{\partial G}{\partial T}\right)_P = -S.$$

A simple substitution yields the equation

$$G = H + T\left(\frac{\partial G}{\partial T}\right)_P \qquad (10.9)$$

Let us now consider two states, distinguished by parameters with subscripts 1 and 2 respectively, and let us apply equation (10.9) to each of them. Then

$$G_1 = H_1 + T\left(\frac{\partial G_1}{\partial T}\right)_P$$

and

$$G_2 = H_2 + T\left(\frac{\partial G_2}{\partial T}\right)_P.$$

On transition from state 1 to state 2, let

$$G_2 - G_1 = \Delta G$$

and

$$H_2 - H_1 = \Delta H.$$

Then

$$\Delta G = \Delta H + T\left(\frac{\partial(\Delta G)}{\partial T}\right)_P.$$

This important relationship is known as the Gibbs–Helmholtz equation. One of its principal uses is in the derivation of the van't Hoff Isochore.

The Gibbs–Helmholtz equation can be re-arranged as follows:-

$$\frac{\Delta G}{T^2} = \frac{\Delta H}{T^2} + \frac{1}{T}\left(\frac{\partial(\Delta G)}{\partial T}\right)_P.$$

Therefore,

$$\frac{1}{T}\left(\frac{\partial(\Delta G)}{\partial T}\right)_P - \frac{\Delta G}{T^2} = -\frac{\Delta H}{T^2}.$$

Now

$$\left(\frac{\partial\left(\frac{\Delta G}{T}\right)}{\partial T}\right)_P = \frac{1}{T}\left(\frac{\partial(\Delta G)}{\partial T}\right)_P - \frac{\Delta G}{T^2}.$$

Therefore,

$$\left(\frac{\partial\left(\frac{\Delta G}{T}\right)}{\partial T}\right)_P = -\frac{\Delta H}{T^2}.$$

This last equation should apply to any transitions including those in which reagents and products are in their standard states. In such a case,

$$\left(\frac{\partial\left(\frac{\Delta G_0}{T}\right)}{\partial T}\right)_P = -\frac{\Delta H}{T^2}.$$

According to the van't Hoff Isotherm, however,

$$\Delta G_0 = -RT \ln K.$$

Hence,

$$\frac{\Delta G_0}{T} = -R \ln K.$$

$$\left(\frac{\partial\left(\frac{\Delta G_0}{T}\right)}{\partial T}\right)_P = -R\left(\frac{\partial(\ln K)}{\partial T}\right)_P$$

$$= -\frac{\Delta H}{T^2}.$$

Therefore,

$$\left(\frac{\partial(\ln K)}{\partial T}\right)_P = \frac{\Delta H}{RT^2}$$

As already explained in Chapter 7, this is the van't Hoff Isochore. It is most commonly used in the integrated form

$$\ln K = \text{constant} - \frac{\Delta H}{RT}.$$

According to this equation, a graph of $\ln K$ against $1/T$ should give a straight line with a negative slope, from which ΔH can readily be determined.

Depression of Freezing Point and Elevation of Boiling Point

Consider a solution at its freezing point. Let the chemical potential of the solvent in the solid phase be μ_{1S}, while that for the liquid phase is μ_{1L}. Otherwise, the same nomenclature will be used as before, a subscript 1 serving for parameters of the solvent, and a subscript 2 for those of the solute.

If the system is at equilibrium at temperature T, then

$$\mu_{1S} = \mu_{1L}$$

and

$$\frac{\mu_{1S}}{T} = \frac{\mu_{1L}}{T}.$$

It therefore follows that any change in either of these two quantities must be accompanied by an equal change in the other. At constant pressure, however, the term on the left of the equation can be affected only by an alteration of temperature, while that on the right is susceptible to changes of both temperature and concentration of the solute. Hence,

$$\left(\frac{\partial\left(\dfrac{\mu_{1S}}{T}\right)}{\partial T}\right)_{P} . dT = \left(\frac{\partial\left(\dfrac{\mu_{1L}}{T}\right)}{\partial T}\right)_{P,N_2} . dT + \left(\frac{\partial\left(\dfrac{\mu_{1L}}{T}\right)}{\partial N_1}\right)_{P,T} . dN_1.$$

According to equation (10.9), however,

$$G = H + T\left(\frac{\partial G}{\partial T}\right)_P .$$

If this is applied to one mole and H is taken as the molar enthalpy,

$$\mu = H + T\left(\frac{\partial \mu}{\partial T}\right)_P ,$$

so that

$$\frac{\mu}{T^2} = \frac{H}{T^2} + \frac{1}{T}\left(\frac{\partial \mu}{\partial T}\right)_P$$

and

$$\left(\frac{\partial\left(\frac{\mu_{1S}}{T}\right)}{\partial T}\right)_P = \frac{1}{T}\left(\frac{\partial \mu_{1S}}{\partial T}\right)_P - \frac{\mu_{1S}}{T^2} = -\frac{H_{1S}}{T^2},$$

where H_{1S} is the molar enthalpy of the solvent in the solid phase. Similarly, if H_{1L} is its molar enthalpy in the liquid phase,

$$\left(\frac{\partial\left(\frac{\mu_{1L}}{T}\right)}{\partial T}\right)_{P,N_2} = \frac{1}{T}\left(\frac{\partial \mu_{1L}}{\partial T}\right)_{P,N_2} - \frac{\mu_{1L}}{T^2} = -\frac{H_{1L}}{T^2}.$$

Also, since

$$\mu_{1L} = \text{constant} + RT \ln N_1,$$

$$\frac{\mu_{1L}}{T} = \frac{\text{constant}}{T} + R \ln N_1.$$

Hence, at constant temperature,

$$\left(\frac{\partial\left(\frac{\mu_{1L}}{T}\right)}{\partial N_1}\right)_{P,T} .dN_1 = \frac{R}{N_1}.dN_1.$$

When these expressions are substituted into the equation giving the total changes of μ_{1S}/T and μ_{1L}/T, the result is

$$-\frac{H_{1S}}{T^2}.dT = -\frac{H_{1L}}{T^2}.dT + \frac{R}{N_1}.dN_1.$$

If the molar latent heat of fusion be denoted by L, then

$$H_{1L} - H_{1S} = L.$$

Also, since

$$N_1 + N_2 = 1,$$

$$dN_1 = -dN_2.$$

An acceptable approximation is provided by the fact that, in a dilute solution,

$$N_1 \approx 1.$$

Accordingly,

$$\frac{\left(H_{1L} - H_{1S}\right)}{T^2}.dT = \frac{R}{N_1}.dN_1.$$

$$\frac{L}{T^2}.dT = -R.dN_2.$$

If the freezing point of the pure solvent is T and that of the solution $(T + \Delta T)$,

$$\int_{T}^{T+\Delta T} \frac{L}{T^2}.dT = -\int_{0}^{N_2} R.dN_2,$$

$$\frac{L}{T} - \frac{L}{T + \Delta T} = -RN_2$$

and

$$\frac{L.\Delta T}{T(T + \Delta T)} = -RN_2.$$

Since ΔT is very small in comparison with T, the approximation equating $T(T + \Delta T)$ with T^2 may once again be made. A very simple re-arrangement then gives the result

$$\Delta T = -\frac{RT^2}{L}.N_2.$$

This gives the same result as before. With slight but obvious modification, the reasoning can be applied to the elevation of boiling point.

The Clapeyron–Clausius Equation

Consider a liquid to be in equilibrium with its vapour. Let the system absorb a very small amount of heat, so that a little of the liquid evaporates reversibly. Let the volume and entropy of the vapour be V and S, respectively, while the corresponding quantities for the liquid are V^* and S^*. Then

increase in Gibbs Free Energy of the vapour $= V.dP - S.dT$

and

increase in Gibbs Free Energy of the liquid $= V^*.dP - S^*.dT$

If the change is a reversible one, however, the Gibbs Free Energy of one mole in the vapour must equal that of one mole in the liquid, and any change in the one must be equal to the simultaneous change in the other. Hence,

$$V.dP - S.dT = V^*.dP - S^*.dT$$

Accordingly,

$$(V - V^*).dP = (S - S^*).dT. \quad (10.10)$$

If one mole is being considered, then V and V^* are molar volumes, and

$$S - S^* = \frac{L}{T},$$

where L is the molar latent heat of evaporation. Equation (10.10) may now be rewritten as

$$\left(V - V^*\right).dP = \frac{L}{T}.dT.$$

Thus,

$$L = T.\frac{dP}{dT}\left(V - V^*\right).$$

This is the Clapeyron–Clausius equation, from which the relationship between vapour pressure and temperature may be derived, for, if the vapour pressures at temperatures T_1 and T_2 are P_1 and P_2 respectively, then for one mole,

$$V - V^* \approx V \approx \frac{RT}{P}.$$

$$L = \frac{RT^2}{P}.\frac{dP}{dT}.$$

$$\int_{P_1}^{P_2}\frac{dP}{P} = \int_{T_1}^{T_2}\frac{L}{R}.\frac{dT}{T^2}.$$

$$\ln\frac{P_2}{P_1} = \frac{L}{R}\left(\frac{1}{T_1} - \frac{1}{T_2}\right).$$

This is the relationship that was previously derived.

A similar argument can be used to describe the relationship between melting point and pressure, but the smaller values of the latent heats cause these two quantities to be less dependent on each other.

Electrochemical Phenomena

The existence of standard electrode potentials can also be deduced by means of Gibbs Free Energy. Consider the following equation:

$$dG = dq - dw_P - dw_e + P.dV + V.dP - T.dS - S.dT,$$

the previous nomenclature being used.
    Then

$$P.dV = dw_P$$

and, if the system is at equilibrium,

$$T.dS = dq.$$

Hence,

$$dG = - dw_e + V.dP - S.dT$$

or, at constant temperature,

$$dG = - dw_e + V.dP. \qquad (10.11)$$

In the case of ions dissolving from a metal rod, P represents the osmotic pressure, so that, if V denotes the volume containing one mole of the ion,

$$V.dP = \frac{RT}{P}.dP.$$

$$\int V.dP = \text{constant} + RT \ln P.$$

Since, however, the osmotic pressure is proportional to the activity, a, of the dissolved ions, the last equation can be written as

$$\int V.dP = \mu_0 + RT \ln a,$$

where μ_0 is a suitable constant. It follows from equation (10.11) that

$$\Delta G = -w_e + \mu_0 + RT \ln a. \quad (10.12)$$

If the change of Gibbs Free Energy refers to one mole of material and the potential of the metal rod is E volts, then

$$w_e = nFE,$$

where n is the valency of the metal and F is the number of coulombs in one faraday. Accordingly, equation (10.12) becomes

$$\Delta G = -nFE + \mu_0 + RT \ln a.$$

At equilibrium, however,

$$\Delta G = 0,$$

so that

$$nFE = \mu_0 + RT \ln a.$$

This can be written as

$$E = \frac{\mu_0}{nF} + \frac{RT}{nF} \ln a,$$

or

$$E = E_0 + \frac{RT}{nF} \ln a.$$

The last statement is the Nernst equation, the implications of which have already been discussed. A slight modification of the argument, which will not be detailed here, leads to the idea of redox potentials.

Conclusion

As can be seen, the concept of free energy may be used to derive all the laws previously obtained from considerations of entropy. The use of free energy is often convenient and sometimes very instructive, but entropy remains the fundamental quantity. When any part of a system is being considered,

$$dS \geq \frac{dq}{T},$$

where the equality refers to a reversible change, and the inequality to an irreversible one. If the whole of an isolated system is under consideration, then

$$dq = 0,$$

so that

$$dS \geq 0.$$

Any other entropy changes are not impossible, but are statistically so unlikely that the possibility of their occurrence can be discounted. This, in its beautiful simplicity, remains the fundamental message of thermodynamics.

Appendix

Partial Differentiation

Consider a quantity z, the value of which is determined by two other quantities denoted by x and y in accordance with the equation

$$z = ax + by. \qquad (A.1)$$

If y is kept constant, then

$$a = \frac{dz}{dx}, \qquad (A.2)$$

and conversely, if x is kept constant,

$$b = \frac{dz}{dy} \qquad (A.3)$$

Differentiation with respect to one variable while the other is kept constant is denoted by the use of the symbol ∂ instead of d, and the variable to be kept constant is often written as a subscript outside a set of brackets. Thus, equations (A.2) and (A.3) may be rewritten as follows:-

$$a = \left(\frac{\partial z}{\partial x} \right)_y$$

$$b = \left(\frac{\partial z}{\partial y} \right)_x.$$

Now if x, y and z are increased by δx, δy and δz respectively, then it follows from equation (A.1) that

$$z + \delta z = a(x + \delta x) + b(y + \delta y). \quad (A.4)$$

By subtraction of (A.1) from (A.4), we obtain

$$\delta z = a.\delta x + b.\delta y.$$

Hence,

$$\delta z = \left(\frac{\partial z}{\partial x}\right)_y .\delta x + \left(\frac{\partial z}{\partial y}\right)_x .\delta y.$$

In the limit, as δx, δy and δz tend towards zero,

$$dz = \left(\frac{\partial z}{\partial x}\right)_y .dx + \left(\frac{\partial z}{\partial y}\right)_x .dy \qquad (A.5)$$

This is the equation that forms the basis of techniques involving partial differentiation, i.e. differentiation with respect to a number of variables one at a time. If x and y are dependent upon some other variable denoted by w, then the rate of change of z with respect to w is given by

$$\frac{dz}{dw} = \left(\frac{\partial z}{\partial x}\right)_y .\left(\frac{\partial x}{\partial w}\right) + \left(\frac{\partial z}{\partial y}\right)_x .\left(\frac{\partial y}{\partial w}\right).$$

Alternatively, equation (A.5) can be used to determine the rate of change of z with respect to x or y as follows:

$$\frac{dz}{dx} = \left(\frac{\partial z}{\partial x}\right)_y + \left(\frac{\partial z}{\partial y}\right)_x .\left(\frac{\partial y}{\partial x}\right).$$

A few other points should be made. The variables x and y must be capable of being changed independently of each other. Although they may be connected via some other variable, neither should have its value determined entirely by the other. Thus, for example, it would be incorrect to write of the internal energy, U, of a gas that

$$dU = \left(\frac{\partial U}{\partial V}\right)_{P,T}.dV + \left(\frac{\partial U}{\partial T}\right)_{P,V}.dT + \left(\frac{\partial U}{\partial P}\right)_{V,T}.dP,$$

since the volume, temperature and pressure of a gas are so connected that stated values of any two of them uniquely determine the third. On the other hand, the equation

$$dU = \left(\frac{\partial U}{\partial V}\right)_{T}.dV + \left(\frac{\partial U}{\partial T}\right)_{V}.dT$$

is perfectly acceptable.

Equation (A.1) refers to the dependence of z upon two variables in a mathematically simple manner. The technique of partial differentiation is, however, perfectly amenable to uses in which the number of independent variables is greater, or the nature of the mathematical relationship is more complicated.

Index